T0291148

Introduction to Steels

Introduction to Steels

Processing, Properties, and Applications

P. C. Angelo
B. Ravisankar

CRC Press is an imprint of the
Taylor & Francis Group, an **informa** business

CRC Press
Taylor & Francis Group
52 Vanderbilt Avenue,
New York, NY 10017

Printed on acid-free paper

International Standard Book Number-13: 978-1-138-38999-1 (Hardback)

Library of Congress Cataloging-in-Publication Data

Names: Angelo, P.C., author. | Ravisankar, B., author.
Title: Introduction to steels : processing, properties, and applications / P. C. Angelo and B. Ravisankar.
Description: New York, NY : CRC Press/Taylor & Francis Group, 2019. | Includes bibliographical references and index.
Identifiers: LCCN 2018052858| ISBN 9781138389991 (hardback : acid-free paper) | ISBN 9780429423598 (ebook)
Subjects: LCSH: Steel–Metallurgy.
Classification: LCC TN730 .A58 2019 | DDC 669/.142–dc23
LC record available at https://lccn.loc.gov/2018052858

Visit the Taylor & Francis Web site at
http://www.taylorandfrancis.com

and the CRC Press Web site at
http://www.crcpress.com

Contents

Preface

Steel is adapting itself to emerging structural standards and conditions. From 1778, when the first iron bridge was built, to the present, steel is the most commonly used material for various applications. Steel is appropriate to most critical requirements by the addition of alloying elements and by suitable heat treatment. Steels are also most suitable for manufacturing by forming, welding, casting, and machining. There are several standard books that deal with the metallurgy of steel but they are meant for metallurgical engineers and are not suitable for engineers from other disciplines including mechanical, production, electrical, electronics, and aeronautical engineering students. It is to fulfill this requirement that *Introduction to Steel: Processing, Properties, and Applications* has been written. It will also be useful for practicing engineers. We also expect the book to satisfy the needs of students preparing for competitive exams.

This book consists of nine chapters covering most of the important types of steels and their physical metallurgy, microstructure, and engineering applications. Chapter 1 introduces the iron-carbon diagram and discusses the importance and interpretation of it as well as the resultant microstructure under equilibrium conditions with respect to carbon content. The effect of nonequilibrium cooling (heat treatment) on properties and methodology of heat treatment are discussed in Chapter 2. The most important surface hardening methods and methodology are described in Chapter 3. Chapter 4 deals with the various plain carbon steels and their heat treatment as well as important applications as engineering material. The effect of alloying elements on phase stability, microstructure, and properties are discussed in Chapter 5. The properties and applications of steel with low amounts of alloying elements are described in Chapter 6. High-strength steels are discussed in Chapter 7, in particular their applications for specialized uses in automobiles and aerospace. Chapter 8 covers high alloy steels – in particular maraging steels, stainless steels, and tool steels – describing their special uses in various applications. The basic techniques of the selection of material illustrated through case studies are included in Chapter 9. In addition, the book contains both subjective and objective questions as an aid to students preparing for semester-end and competitive exams.

This book is based on our many years of teaching undergraduate and postgraduate students of various engineering disciplines. It is our fond hope that the present book will serve as a textbook for courses in engineering and also as a useful reference book for practicing engineers. We look forward to receiving your valuable response and suggestions.

Author Bios

P.C. Angelo, PhD, is internationally renowned in the fields of Powder Metallurgy, Materials Characterisation for over 40 years. He has taught, researched, and published extensively in the areas of optical emission spectroscopy, optical microscopy, X-ray diffraction, X-ray fluorescence, electron probe microanalysis, scanning electron microscopy, and scanning Auger microprobe. Dr. Angelo began his career in 1962 as a scientist at the Defence Metallurgical Research Laboratory (DMRL), Hyderabad, India, one of the laboratories of Defence Research and Development Organization (DRDO), Government of India, New Delhi. After 35 years of service he retired from DRDO and joined PSG College of Technology in the Department of Metallurgical Engineering and worked as Director, Metal Testing and Research Centre and as Professor and Dean for 20 years and retired in 2017. His books *Powder Metallurgy: Science, Technology and Applications*, *Materials Characterisation*, and *Non Ferrous Alloys* have been well received as textbooks by numerous institutions. Presently, he is Visiting Professor at PSG College of Technology, Coimbatore, Tamil Nadu, India.

Dr. B. Ravisankar, Professor, Department of Metallurgical and Materials Engineering, National Institute of Technology, Tiruchirappalli, has more than 25 years of teaching-cum-research experience. He has contributed a chapter on Equal Channel Angular Pressing (ECAP) to the *Handbook of Mechanical Nanostructuring*. He also published a book on *Non Ferrous Alloys*.

1

Iron-Carbon Diagram

1.0 Introduction

Steel is frequently the "gold-standard" even among the emerging structural materials. Steels have numerous uses. Steel is used more than any other metal for producing alloys. In 1778 the first iron bridge was built. In 1788, iron water pipe lines were laid. In 1818 the first steel ship was launched. In 1889 Gustav Eiffel, a French engineer, built the Eiffel Tower with steel. Eiffel's contemporaries thought that his 300-meter latticed structure would prove too fragile to last. But Eiffel argued that his creation would stand at least for a quarter of a century. Even today the Eiffel Tower – the true landmark of Paris – is intact and attracting visitors from all over the world.

Steel is a moving standard, since regular and exciting discoveries are being made. This makes steel to remain the most successful and cost-effective of all materials ever. Major reason for the overwhelming dominance of steel is the variety of microstructures and properties that can be generated by solid-state transformation and processing. Therefore, in studying advanced steels, it is useful to discuss, first the nature and behavior of pure iron, then the iron-carbon alloys, and finally the complexities that arise when further solutes are added.

At least four allotropes of iron occur naturally in bulk form: body-centred cubic (bcc, α and δ, ferrite), face-centred cubic (fcc, γ, austenite) and hexagonal close packed (hcp, ε). As molten iron cools past its freezing point of 1538°C, it crystallizes into a body-centered cubic (bcc) δ allotrope. As it cools further to 1394°C, it changes into a face-centered cubic (fcc) γ-iron allotrope known as austenite. At 912°C and below, the crystal structure again becomes bcc α-iron allotrope, or ferrite. At pressures above approximately 10 GPa and temperatures of a few hundred Kelvin or less, α-iron changes into a hexagonal close-packed (hcp) structure that is also known as ε-iron; the higher-temperature γ-phase also changes into ε-iron, but does so at still higher pressure. The phase diagram for pure iron is illustrated in Figure 1.1.

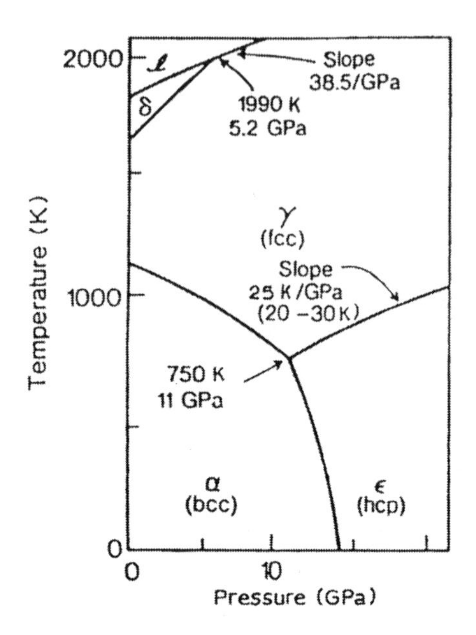

FIGURE 1.1
The phase diagram for pure iron

The phase β in the alphabetical sequence α, β, γ, δ … is missing because the magnetic transition in ferrite was at one time incorrectly thought to be the β allotrope of iron. In fact, there are magnetic transitions in all allotropes of iron.

Some controversial experimental evidence exists for a fifth natural allotrope of iron in the core of the earth, where the pressure reaches some three million times that at the surface and where the temperature is estimated to be about 6000°C. The core of the earth is predominantly iron, and consists of a solid inner core surrounded by a liquid outer core. Knowledge of the core is uncertain, but it has been suggested that the crystal structure of the solid core may be an orthorhombic or double hcp, also denoted as β (not to be confused with β named for magnetic transition for ferrite). Calculations that assume pure iron, indicate that the ε iron remains the most stable under inner-core conditions. These high-pressure phases of iron are of no practical importance, but are important as the end member models for the solid parts of planetary cores.

Pure iron (99.99%) is not an easy material to produce and also, iron of high purity is extremely weak; the resolved shear stress of a single crystal of iron at room temperature is as low as 10 MPa, while the yield stress of a polycrystalline sample of iron at the same temperature is well below 50 MPa and has a hardness of 20–30 Brinell. But pure iron has nevertheless been made with a total impurity content of less than

60 parts per million (ppm), of which 10 ppm is accounted for by nonmetallic impurities such as carbon, oxygen, sulfur and phosphorus, with the remainder representing metallic impurities. With the controlled amount of nonmetallic impurities especially carbon (between 0.002% and 2.1%) produces steel that may be up to 1000 times harder than pure iron. Maximum hardness of 65 HRc is achieved with 0.6% carbon content. The possibility of achieving a variety of microstructure due to the allotropic behavior of iron, good workability (deformability) due to the softness of iron and the strength achieved due to carbon, make the steel suitable for a wide variety of applications. Steel is an enigma as it rusts easily, yet it is the most important of all metals. Steel is 90% of all metals being refined today as steel is useful in terms of its mechanical, physical and chemical properties.

It is well known that steel is an interstitial solid solution (alloy) of iron and carbon and it is more often referred to as a metal and while with subsequent addition of other solutes it is called as alloy steel, though it is a misnomer. Alloy steels are by far the most common industrial metals as they have a great range of desirable properties. Steel, with smaller carbon content than pig iron (about 4 wt.% C) but more than wrought iron (almost a pure iron), was first produced in antiquity by using a bloomery. By 1000 BCE, blacksmiths in Luristan in western Persia were making good steel. The improved versions such as Wootz steel by India and Damascus steel were developed around 300 BCE. These methods were specialized, and so steel did not become a major commodity until 1850s. New methods of producing it by carburizing bars of iron in the cementation process were devised in the 17th century. In the Industrial Revolution, new methods of producing bar iron without charcoal were devised and these were later applied to produce steel. In the late 1850s, Henry Bessemer invented a new steelmaking process, involving blowing air through molten pig iron, to produce mild steel. This made steel much more economical, thereby leading to wrought iron being no longer produced in large quantities. Carbon steels are least expensive of all metals while stainless steels are costly.

1.1 Iron-Carbon Diagram

Of all the binary alloy systems, the one that is possibly the most important is iron and carbon. Both steels and cast irons, primary structural materials in every technologically advanced culture, are essentially iron–carbon alloys. This section deals with the study of the phase diagram and the solidification and development of several of the possible microstructures in iron-carbon alloys.

Iron–carbon phase diagram is presented in Figure 1.2. Pure iron, upon heating, experiences two changes in crystal structure before it melts. At room temperature the stable form, called alpha ferrite, or iron, has a BCC crystal structure. Ferrite experiences a polymorphic transformation to FCC austenite, or iron, at 912°C. This austenite persists up to 1394°C; at that temperature the FCC austenite reverts back to a BCC phase known as delta ferrite that finally melts at 1538°C. All these changes are shown along the left vertical axis of the phase diagram.

The composition axis in Figure 1.2 extends only to 6.70 % C as at this concentration the intermediate compound iron carbide, or cementite (Fe_3C), is formed that is represented by a vertical line on the phase diagram. Thus, the iron-carbon system may be divided into two parts: an iron-rich portion, as in Figure 1.2; and the other (not shown) for compositions between 6.70 and 100% C (pure graphite). In practice, all steels and cast irons have carbon contents less than 6.70% C and therefore, only the iron–iron carbide system is considered. Figure 1.2 should be more appropriately labeled as Fe–Fe_3C phase diagram, as Fe_3C is now considered to be a component. Convention and convenience dictate that composition be

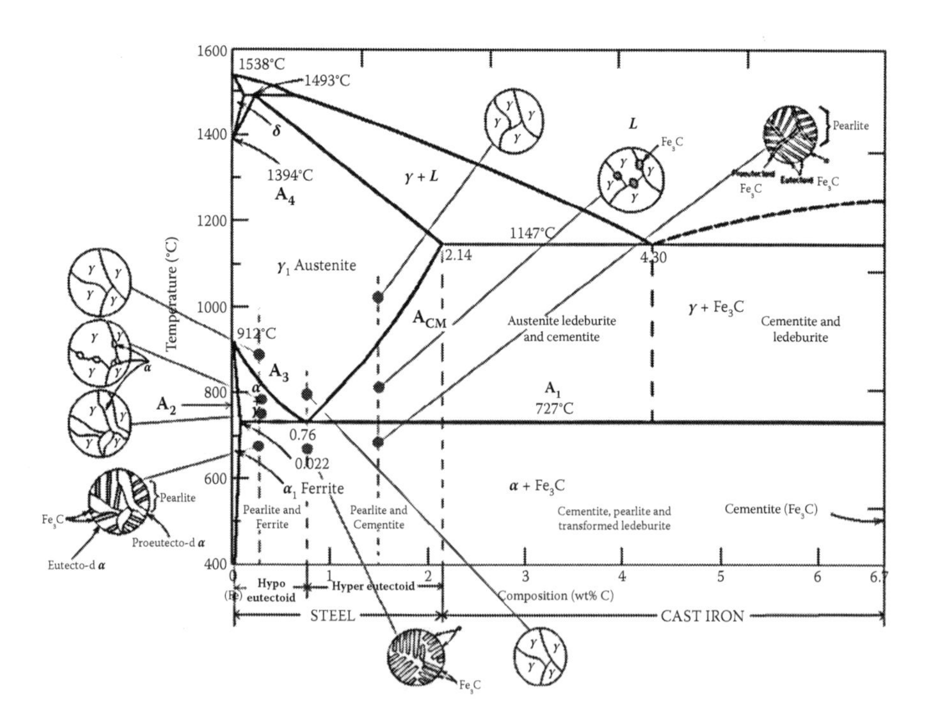

FIGURE 1.2
Iron-carbon (iron-iron carbide) phase diagram

expressed in "wt% C" rather than "wt% Fe_3C"; 6.70 wt% C corresponds to 100 wt% Fe_3C. In this book, all percentages refer only to wt% unless otherwise mentioned.

Carbon is an interstitial impurity in iron and forms a solid solution with ferrites, and also with austenite, as indicated in Figure 1.2. In the BCC ferrite, only small concentrations of carbon are soluble; the maximum solubility is 0.022% at 727°C. The limited solubility is explained by the shape and size of the BCC interstitial positions that make it difficult to accommodate the carbon atoms. Even though present in relatively low concentrations, carbon significantly influences the mechanical properties of ferrite. This particular iron–carbon phase is relatively soft, may be made magnetic at temperatures below 768°C. Figure 1.3 (a) is a photomicrograph of ferrite.

The austenite, or gamma phase of iron, which is nonmagnetic, when alloyed with just carbon, is not stable below 727°C, as indicated in Figure 1.2. The maximum solubility of 2.14% carbon in austenite, occurs at 1147°C. This solubility is approximately 100 times greater than the maximum for BCC ferrite, as the FCC interstitial positions are larger, and therefore, the strains imposed on the surrounding iron atoms are much lower. Also the phase transformations involving austenite are very important in the heat treating of steels. Figure 1.3 (b) shows a photomicrograph of austenite phase.

The delta ferrite is virtually the same as alpha ferrite, except for the range of temperatures over which each exists. Because the delta ferrite is stable only at relatively high temperatures, it is of no technological importance and hence is not discussed further.

Cementite (Fe_3C) forms when the solubility limit of carbon in ferrite is exceeded below 727°C (for compositions within the Fe_3C phase region). As

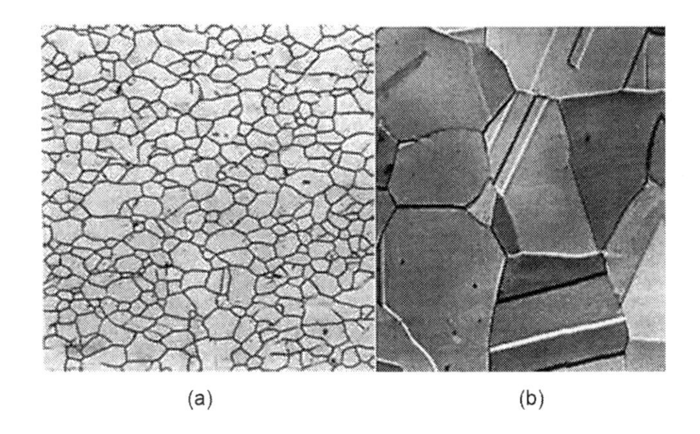

FIGURE 1.3
Photomicrograph of (a) ferrite and (b) austenite

indicated in Figure 1.2, Fe_3C will also coexist with the phase between 727 and 1147°C. Cementite is very hard and brittle; the strength of some steels is greatly enhanced by its presence. Strictly speaking, cementite is only metastable; that is, it will remain as a compound indefinitely at room temperature. But if heated to between 650 and 700°C for several years, it will gradually change or transform into iron and carbon in the form of graphite that will remain upon subsequent cooling to room temperature. Thus, the phase diagram in Figure 1.2 is not a true equilibrium one because cementite is not an equilibrium compound. However, in as much as the decomposition rate of cementite is extremely sluggish, virtually all the carbon in steel will be as Fe_3C instead of graphite, and the iron–iron carbide phase diagram is, for all practical purposes, valid.

According to the iron-carbon diagram, alloys with less than 2% carbon are known as steels and alloys with more than 2% carbon are cast irons. There are three invariant (phase) reactions occurring in the iron-carbon system:

(i) A peritectic reaction at 1493°C at 0.16% carbon

delta ferrite + Liquid ↔ austenite
(0.08%C) (0.5%C) (0.18%C)

This reaction is of minor importance in steels although it has some importance in welding of austenitic stainless steels, which will be discussed later in this book.

(ii) The second one is the eutectic reaction, at 4.30% C and 1147°C;

Liquid ↔ austenite + cementite
(4.3%C) (2%C) (6.67%C)

Subsequent cooling to room temperature will promote additional phase changes. This reaction has significant impact in cast irons. The eutectic mixture is also called as ledeburite.

(iii) The third is eutectoid invariant reaction at a composition of 0.76%C at a temperature of 727°C.

Austenite ↔ ferrite + cementite
(0.76%C) (0.025%C) (6.67%C)

The eutectoid phase changes are very important, being fundamental to the heat treatment of steels, as explained in subsequent discussions. The eutectoid mixture is called pearlite.

Although a steel alloy may contain as much as 2.0%C, in practice, carbon concentrations rarely exceed 1.0%.

Various microstructures that develop depend on both the carbon content and heat treatment (cooling rate and pattern). Initially, the discussion is confined to very slow cooling of steel alloys, in which equilibrium is continuously maintained. A more detailed exploration of the influence of heat treatment on microstructure, and ultimately on the mechanical properties of steels, is discussed subsequently.

1.2 Slow Cooling of Eutectoid Composition

Phase changes that occur upon passing from the austenite region into the ferrite + cementite phase field are relatively complex. An alloy of eutectoid composition cooled from a temperature within the austenite phase region, above eutectoid temperature (Figure 1.4) and moving down the vertical line R is initially composed entirely of the austenite phase having a composition of 0.76%C and Figure 1.4 shows the corresponding microstructure. As the alloy is cooled, no changes will occur until the eutectoid temperature (727°C) is reached. Upon crossing this temperature to point E, the austenite transforms into ferrite and cementite (eutectoid reaction).

The microstructure for eutectoid steel that is slowly cooled through the eutectoid temperature consists of alternating layers or lamellae of the two

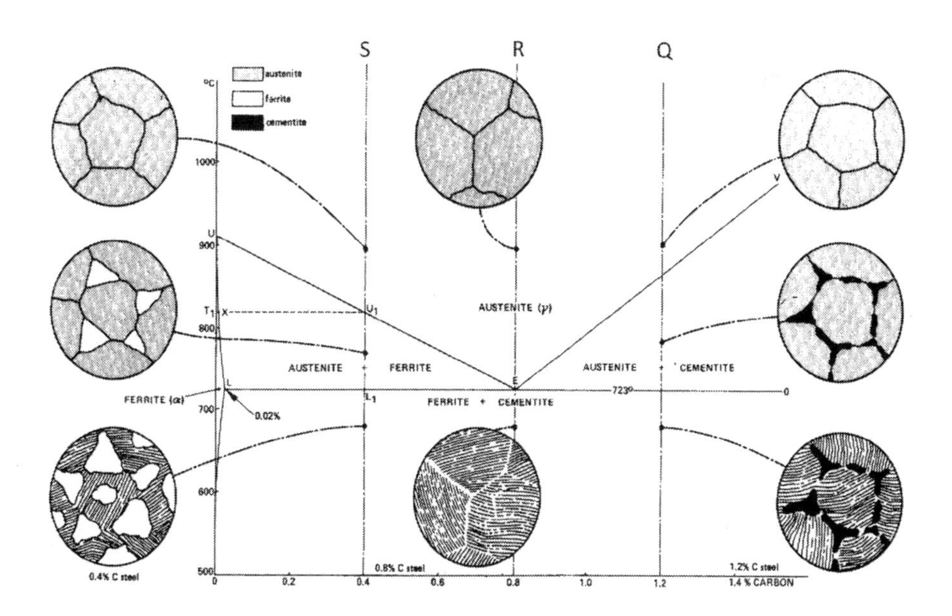

FIGURE 1.4
Microstructure developed during slow cooling of various steels

phases (ferrite and Fe_3C) that form simultaneously during the transformation. This microstructure, represented schematically in Figure 1.4, point E, is called pearlite because it has the appearance of mother of pearl when viewed under the microscope at low magnifications. The pearlite exists as grains, often termed "colonies" and within each colony the layers are oriented essentially in the same direction, while varying from one colony to another. Pearlite has properties intermediate between the soft, ductile ferrite and the hard, brittle cementite.

1.3 Slow Cooling of Hypoeutectoid Composition

Left of the eutectoid, between 0.022 and 0.76%C is termed as hypoeutectoid (less than eutectoid) alloy. Let us consider slow cooling an alloy of composition of 0.4%C represented by moving down the vertical line S in Figure 1.4. At about 875°C, the microstructure will consist entirely of grains of the austenite phase, as shown schematically in Figure 1.4. Cooling to about 775°C, which is within the ferrite + austenite phase region, both of these phases will coexist as in the schematic microstructure. Most of the small particles will form along the original austenite grain boundaries. While cooling an alloy through the ferrite + austenite phase region, the composition of the ferrite phase changes with temperature along the ferrite – (ferrite + austenite) phase boundary (line UL) becoming slightly richer in carbon. On the other hand, the change in composition of the austenite is more dramatic, proceeding along the (ferrite + austenite) – ferrite boundary (line UE) as the temperature is reduced.

Cooling from just above the eutectoid but still in the ferrite + austenite region, will produce an increased fraction of the phase and a microstructure similar to that shown: the particles would have grown larger. At this point, the composition of the ferrite phase will contain 0.022%C, while the austenite phase will be of the eutectoid composition, 0.76%C. As the temperature is lowered just below the eutectoid, all the austenite phase present will transform to pearlite, according to the eutectoid reaction. There will be virtually no change in the phase after crossing the eutectoid temperature and the microstructure will appear as schematically shown in Figure 1.4. Thus the ferrite phase will be present both in the pearlite (eutectoid mixture) and also as the phase that formed while cooling through the ferrite + austenite phase region. The ferrite that is present in the pearlite is called eutectoid ferrite, whereas the other, that formed above eutectoid temperature, is termed proeutectoid (meaning pre- or before eutectoid) ferrite.

The amount of proeutectoid ferrite and pearlite can be determined by drawing a tie line between the compositions of 0.008% carbon (almost left axis of the iron carbon diagram) and 0.76% carbon. Alternatively the amount of phases can also be found by simple thumb rule namely, the

0.8% carbon steel contains 100% pearlite and so 0.4% carbon steel will contain about 50% pearlite and rest 50% proeutectoid ferrite. To find the total ferrite content, a tie line should be drawn between the solubility limit of carbon in ferrite and to the right vertical axis of the iron- carbon diagram, that is, 6.67% carbon or 100% cementite.

1.4 Slow Cooling of Hypereutectoid Composition

Hypereutectoid alloys are those containing between 0.76 and 2.14% carbon, which are cooled from temperatures within the austenite phase field. Consider an alloy of composition of 1.2% carbon in Figure 1.4, which, upon cooling, moves down the line Q. Above line EV only the austenite phase will be present with a composition of 1.2% carbon; the microstructure will appear as shown, having only austenite grains. Upon cooling into the austenite + cementite phase field, the cementite phase will begin to form along the initial austenite grain boundaries, similar to the phase in Figure 1.4. This cementite is called proeutectoid cementite – that forms before the eutectoid reaction. Of course, the cementite composition remains constant (6.70%C) as the temperature changes. However, the composition of the austenite phase will move along line EV toward the eutectoid. As the temperature is lowered through the eutectoid point, all remaining austenite of eutectoid composition is converted into pearlite; thus, the resulting microstructure consists of pearlite and proeutectoid cementite as microconstituents (Figure 1.4).

In the photomicrograph of a 1.2% carbon steel, note that the proeutectoid cementite appears light. Because it has much the same appearance as proeutectoid ferrite, there is some difficulty in distinguishing between hypoeutectoid and hypereutectoid steels on the basis of microstructure.

During the nonequilibrium cooling of the iron-carbon alloys, conditions of metastable equilibrium have been maintained continuously; that is, sufficient time has been allowed at each new temperature for necessary adjustment in phase composition and relative amounts of microconstituents predicted from the iron-carbon diagram. The microstructure developed due to the nonequilibrium cooling plays major role in determining the mechanical properties of the steels. The two nonequilibrium effects of practical importance are:

(i) the occurrence of phase changes or transformations at temperatures other than those predicted by phase boundary lines on the phase diagram

(ii) the existence at room temperature of nonequilibrium phases that do not appear on the phase diagram.

Both are discussed in Chapter 2.

2

Heat Treatment of Steels

2.0 Introduction

The transformation of carbon steel from one microstructure or crystalline structure to another makes the material heat treatable. In other words, it allows for changes in the properties of the material just by going through various heating and cooling cycles, without a change in the overall chemical composition of the material. This characteristic can also result in property changes during fabrication processes such as hot bending/forming, welding, and brazing.

Heat treatment of steel is defined as the process of heating and cooling steel to improve properties such as toughness, machinability, hardness, ductility, as well as to remove residual stresses, wear resistance, grain refinement, and so on. Basically, there are four types of heat treatments: (i) annealing, (ii) normalizing, (iii) hardening, and (iv) tempering. There is one more heat treatment process called precipitation hardening. Although precipitation hardening has minor role in plain carbon steels and low alloy steels, it plays a major role in strengthening high alloy steels such as stainless steels, maraging steels, and so on. For the sake of completion, precipitation hardening is also briefly described. Heat treatment also covers surface hardening methods such as flame hardening, induction hardening, carburizing, nitriding, cyaniding, carbo nitriding, and so on, which are dealt in next chapter.

The material specifications provide the specific heat treatments to achieve the properties for the specific material or application. Heat treatment is highly dependent on the manufacturing methods used for that product and the requirements can range from no heat treatment to subcritical heat treatments (such as precipitation heat treatment, tempering, or stress relief) to high-temperature (austenitizing) heat treatments (such as quench hardening, annealing, or normalizing) that might be followed by a tempering heat treatment. Descriptions of common heat treatments are given in the following sections.

2.1 Annealing

Annealing is a very broad term used to describe a variety of heat treatments and it is a process customarily applied to remove stresses or work hardening. For the purpose of heat treatment used on carbon steels in the material specifications, the more specific term "full annealing" better describes the process. Full annealing is defined as "annealing a steel object by austenitizing it and then cooling it slowly through the transformation range." This results in maximum transformation to ferrite and to coarse pearlite giving the lowest hardness and strength. Full annealing of carbon steels would require the material to be heated to austenitizing temperature which must be determined from the iron-carbon diagram. Generally, the austenitizing temperature for all practical purposes is 50°C above the austenitizing temperature determined from the iron carbon diagram to have a homogenous structure and to nullify the effect of heating rate. Due to annealing, there is no change in microstructure or amount of microconstituents. They differ only in morphology and distribution of the phases. The final microstructure will be the same as that of slow cooled conditions. Annealing is applied to forgings, cold-worked sheets and wires, welded parts as well as castings.

The three main annealing processes used for steels are described here.

1. ***Subcritical Annealing***

Subcritical annealing is carried out below A1 temperature and there is no austenite transformation involved. The main aim is to relieve residual stresses caused by previous processing such as cold working, welding, or casting (stress relief anneal) or to recrystallize cold-worked material. The sub critical annealing is also referred to as stress relieving annealing. Stress relieving is defined as "heating a steel object to a suitable temperature, holding it long enough to reduce residual stresses, and then cooling it slowly enough to minimize the development of new residual stresses." Locked in (residual) stresses in a component cannot exist at a greater level than the yield strength of the material. An increase in the temperature of steel lowers the yield strength and thus relieves some of the stresses. Further reduction in the residual stress can occur due to a creep mechanism at high stress relief temperatures. Stress relieving has a time-temperature relationship. Although some stress relief occurs very quickly as a result of the lower yield strength at temperature, additional stress relief occurs by the primary creep mechanism. Stress relief temperatures are typically 595–675°C for carbon steels. Stress relieving annealing temperature is lower than annealing temperatures.

The other subclass of subcritical annealing is process annealing. It is commonly employed for wires and sheets and carried out between 500° to 690°C for several hours. Process anneal is carried out at higher temperatures

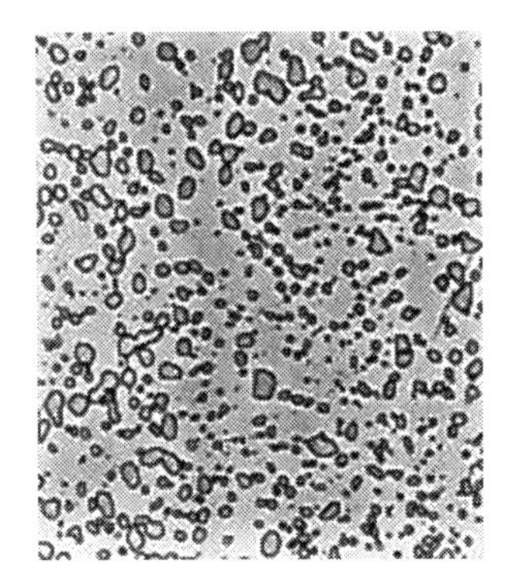

FIGURE 2.1
Spherodized microstructure

(usually 11° to 22°C below A1) than stress relief anneal (usually between 500° to 650°C).

2. ***Spherodise Annealing***

Spherodise annealing is carried out between A1 and A3 for hypoeutectoid steels or between A1 and Acm for hypereutectoid steels. Steel is heated to within 55°C above A1 and then transformed at a temperature less than 55°C below A1 to produce a structure consisting of spherodized carbide particles in a ferrite matrix (Figure 2.1). Alternatively, the steel can be held at a temperature just below A1 for sufficient time for the cementite lamellae of the pearlite to spherodize. This happens because it leads to a reduction in surface energy of the cementite-ferrite interfaces. The spherodized structure (spheroidite) has the minimum hardness, the maximum ductility, as well as the maximum machinability especially in higher carbon pearlitic steels.

3. ***Supercritical or Full Annealing***

Supercritical or Full annealing is carried out above A3 for hypoeutectoid steels and between A1 and Acm for hypereutectoid steels. In full annealing, there is transformation to austenite on heating and back to ferrite and pearlite upon cooling.

2.2 Normalizing

Normalizing is a specific term defined as "heating a steel object to a suitable temperature above the transformation range and then cooling it

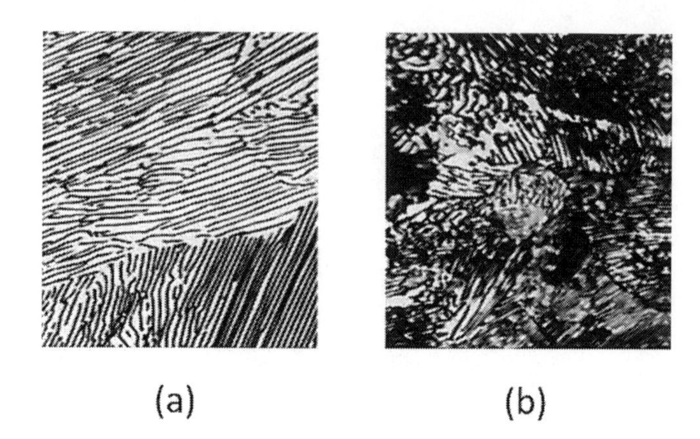

FIGURE 2.2
Pearlite in (a) annealed and (b) normalized conditions.

in air to a temperature substantially below the transformation range." Typical normalizing temperatures are 55°C above A3 for hypoeutectoid steels and 27°C above Acm for hypereutectoid steels. The air cooling may be natural or slightly forced air convection based on the dimensions of the

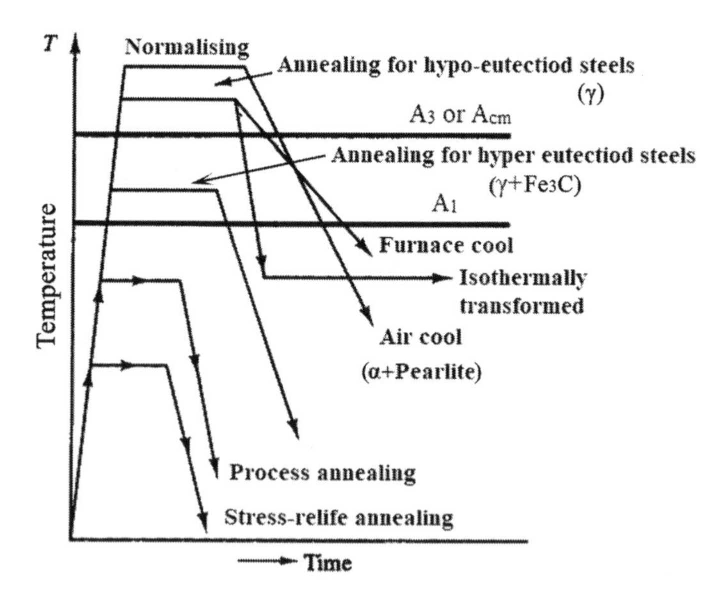

FIGURE 2.3
Heating and cooling cycles for stress relief, process, full and normalizing of hypo- and hypereutectoid steels

steel billet. The high cooling rate (cooling in air) results in ferrite and pearlite but the phase fraction changes. There is no sufficient time for the precipitation of proeutectoid ferrite and hence amount of fine pearlite is more. For example, 0.4% carbon steel with 50% ferrite and 50% pearlite when normalized will have about 65% fine pearlite and 35% ferrite. And so the strength of the normalized steel is higher than annealed steels. In addition, the morphology of pearlite in the annealed and normalized conditions differs. The lamellar structure of ferrite and cementite is fine in normalized pearlite as compared to annealed pearlite as shown in Figure 2.2.

It is often used after hot rolling, where a high finishing temperature can lead to a coarse microstructure. Benefits of normalizing include improved machinability, grain size refinement, homogenization of composition and modification of residual stresses. Figure 2.3 represents various types of annealing and normalizing for hypo and hypereutectoid steels.

2.3 Hardening

"Hardening a steel object is by austenitizing it and then cooling it rapidly enough so that some or all of the austenite transforms to martensite." The rapid cooling suppresses the formation of austenite to $\alpha + Fe_3C$, instead it produces hard martensite for applications that demand a hard material (e.g., knives, razor blades, surgery tools, cutting tools, etc.). Only the austenite transforms to martensite. If austenite + ferrite is quenched, all of the austenite transforms completely to martensite, while the ferrite remains unchanged. Similarly, when austenite + cementite is quenched, again all of the austenite completely transforms to martensite while the cementite remains unchanged. Hence, complete austenitizing is the important step in hardening process.

The hardening is a nonequilibrium cooling process and so the phase formed due to hardening (martensite) will not be found in the iron-carbon diagram since they are plotted in equilibrium cooling conditions. Hence, a scheme for understanding the transformations with respect to time is called transformation diagrams. Transformation diagrams were first published by Bain and Davenport in the United States in 1930 paving the way to the detailed understanding of nonequilibrium cooling, isothermal cooling and the effects of alloying elements on the heat treatment response in steels.

2.3.1 Transformation Diagrams

There are several different variations of transformation diagrams while the most commonly used and referenced are the isothermal transformation

(IT) diagram (commonly called as time-temperature-transformation(TTT) diagram) and the continuous cooling transformation (CCT) diagram. All the transformation diagrams plot temperature versus log time, with the display showing the expected crystalline structures and microstructures. The IT diagram shows the expected result when the steel is held for varying lengths of time, keeping the temperature essentially constant (after an initial austenization). The CCT diagram shows the expected result when the steel is cooled continuously at varying rates from the austenitic phase. These transformation diagrams appear to be similar, but the CCT transformation curves are typically depressed and moved to the right during continuous cooling from those in the IT diagram. The IT diagrams are more commonly used in predicting structures during heat treatment while CCT diagrams are used during welding and casting of steels.

The isothermal diagrams (IT) are plotted as follows: A steel is first heated to a temperature in the austenitic range, typically 20°C above the A1 or Acm line, and then cooled rapidly in a bath to a lower temperature, allowing isothermal transformation to proceed. The progress of transformation can be followed by dilatometry, the degree of transformation depending upon the holding time at the temperature. Figure 2.4 illustrates schematically the start and finish of transformation to ferrite, pearlite, bainite, and martensite on a diagram as a function of temperature and time.

Figure 2.4 is a schematic IT diagram for carbon steel. The transformation curve shows transformation with respect to time (rate of cooling). The cooling curves (line 1 to 7) are superimposed for self-explanation. When the cooling rate is fast (line 6 and 7), it ends up with martensite structure once it is cooled below Mf (martensite finish) temperature. The caution is that the cooling rate line should not touch the beginning of the softer products such as ferrite and pearlite. The critical cooling rate (CCR) is important to achieve full martensitic transformation. The cooling curve (line 7) represents the critical cooling rate. When cooling rate is slightly lower than critical cooling rate (line 5), the softer products start precipitating but the cooling is fast enough so that the transformation is not completed. It results in partial softer products and martensite upon cooling below martensite finish temperatures. On further decreasing the cooling rate (lines 1 to 4), the transformation of austenite to softer product is completed and never ends up with martensite, which is annealing or normalizing. The IT diagrams are drawn for a fixed composition. The addition of carbon or other alloying elements alter the diagrams. The effect is discussed further in later sections.

During hardening, the cooling rate is too rapid to allow nucleation and growth mechanisms (critical cooling rate) and the result is that the trapped carbon is forced into the crystalline lattice. Instead of forming ferrite

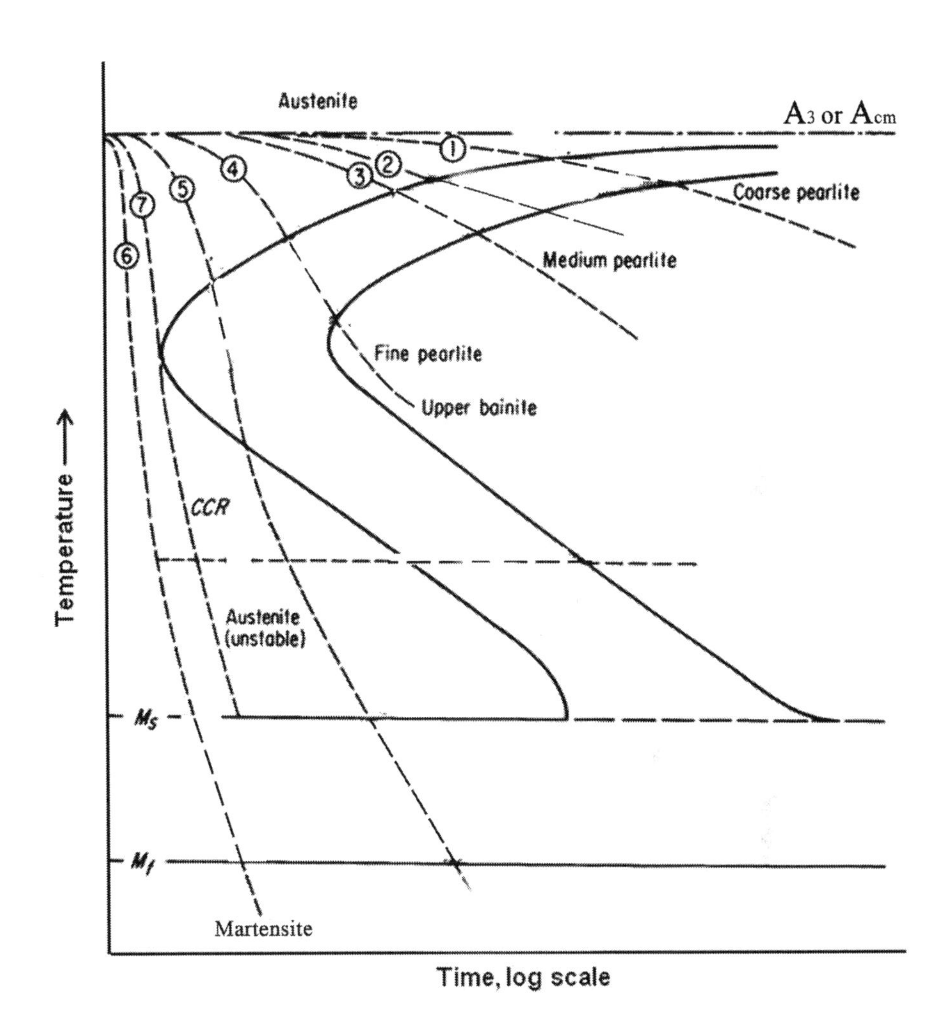

FIGURE 2.4
IT (TTT) diagram for a carbon steel superimposed with cooling curves (Schematic)

structures, the austenite lattice shears and results in a body-centered tetragonal structure called martensite. This martensitic transformation occurs without diffusion of the carbon and therefore occurs very rapidly. In addition, once the austenitic structure is undercooled to the point at which the carbon cannot diffuse and additional ferrite cannot form, the only remaining transformation that can occur upon further cooling is to martensite. The temperature at which martensite begins to form from austenite is the martensite start (Ms) temperature. Because ferrite cannot form, martensite will continue to form as the temperature decreases from any existing austenite until all of the austenite is transformed, which occurs at the martensite finish (Mf) temperature. This carbon steel martensitic structure is known to be both hard and strong but lacks ductility and toughness in the untempered state. There are two types of martensite; Lath and Plate. The formation of the type of martensite depends on the carbon content. Lath type martensite is formed when carbon content is low while plate type martensite is formed when carbon content in steels is higher. The resulting maximum hardness is closely related to the carbon content of the steel and the percentage of martensite formed.

2.4 Hardenability

Hardenability is defined as the depth to which the hard martensite is able to form upon quenching. Since for an actual component it depends upon the thickness, the cooling rate varies across the cross section (Figure 2.5). The surface cools fast upon quenching and further moving toward the center the cooling rate decreases in different microphases across the cross section.

The hardenability of steels is measured by various methods; the Jominy end quench test is the most popular. Figure 2.6 schematically represents the Jominy end quench test as per ASTM standard A 255. The specimen is heated to the austenitizing temperature and fixed in the holder. A water jet is forced from the bottom end. The sample experiences different cooling rate along its length. After cooling, the sample is taken from the holder and its hardness is measured along its length from quenched end to the other end.

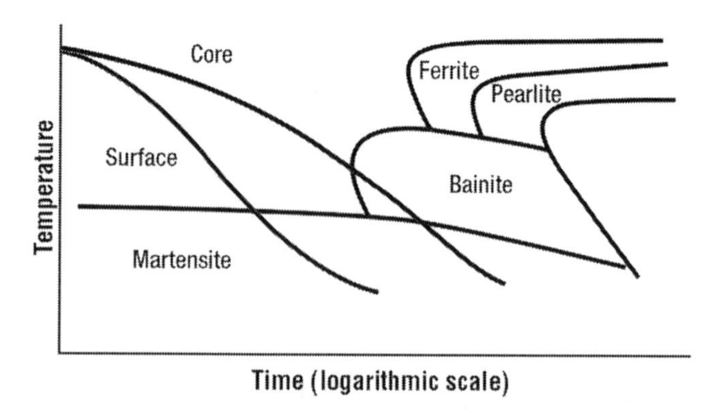

FIGURE 2.5
Variation in cooling rate between surface and core superimposed on IT (TTT) diagram (Schematic)

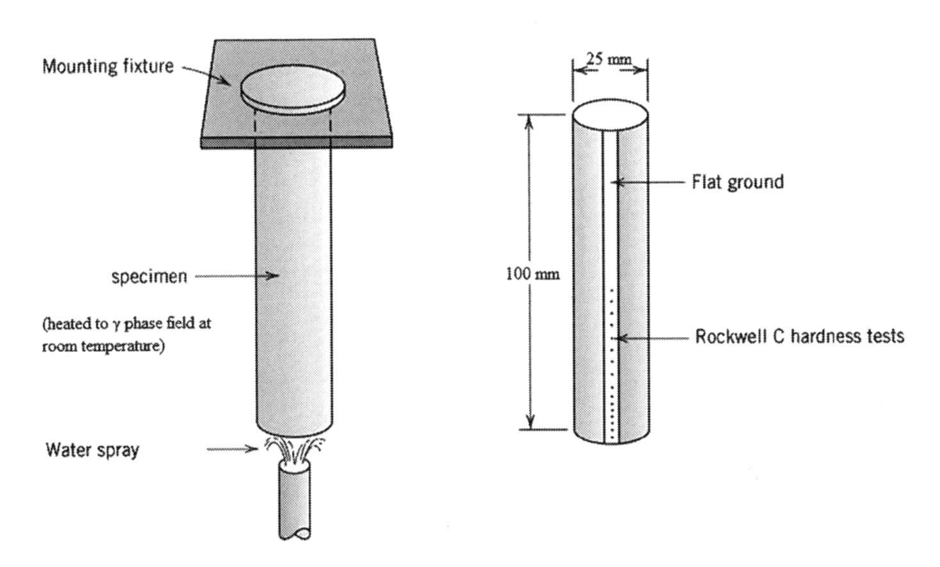

FIGURE 2.6
Schematic representation of the Jominy end quench test

The Jominy end-quench test is related to the CCT diagram because the specimen is raised to an austenitizing temperature and then quenched from one end with water. The result is a varying cooling rate along the specimen length that can then be plotted on the CCT to determine the expected microstructure. Determination of the microstructure is a bit tedious process and as the microstructure obtained is directly related to the hardness which is easy to measure. Hence a graph plotted between the distance from the quenched end to the other end against hardness is used to determine the hardenability of the steel.

The critical cooling rate decides the depth to which the component can be hardened which in turn is decided by the IT (TTT) diagram of the steel. If the beginning of transformation of softer products (commonly referred as nose of the C curve) is moved toward right (away from the Y-axis-Figure 2.4), the critical cooling rate to achieve martensite is low and hence the portion which experiences lower cooling rate can also form martensite. This results in a steel with higher hardenability. Generally alloying elements added to steel increase the hardenability, as shown in Figure 2.7.

By contrast, if the nose of the C curve is moved toward the left (towards Y axis-Figure 2.4), the critical cooling rate is high to achieve martensite. Such a high cooling rate is almost impossible to achieve in practical conditions. A typical example is the plain carbon steels with carbon less than 0.25%. In practical conditions, minimum of 0.3% carbon is required to achieve martensite on the surface.

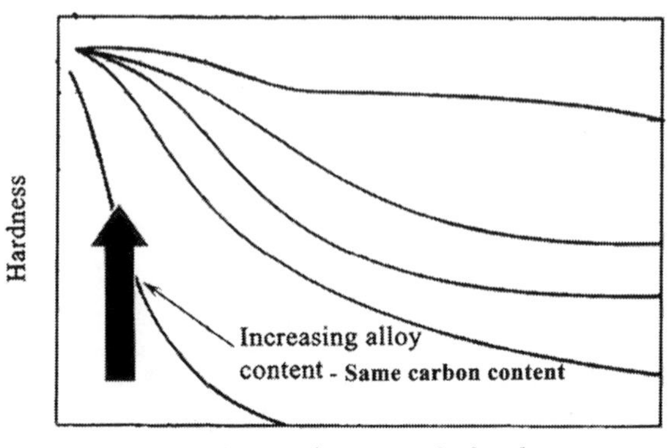

FIGURE 2.7
Effect of alloying elements on hardenability (Schematic)

Apart from water, a range of quenchants are used. The order of severity is from 5% caustic soda, 5–10% brine, cold water, warm water, mineral oil, animal oil, and vegetable oil. When the order of severity is high (5% caustic soda), the cooling rate is faster. As mentioned in the previous paragraph, steels with less than 0.25%C cannot be hardened by quenching even in 5% caustic soda. The quenching medium is selected based on the C curve (Figure 2.4) or the critical cooling rate. Care should be taken in the selection of the quenchants since if the cooling rate is lower than critical cooling rate, martensite may not form for adequate depth and if the cooling rate is more, possibility of quench cracks will be more due to high thermal stresses.

2.5 Tempering

Quench hardening is normally the first step in a heat treatment which would then include a subsequent tempering heat treatment. The martensitic steel is excessively hard and strong with characteristic low toughness and hence the tempering treatment is used to recover some of the more desirable properties. Tempering is defined as "reheating hardened steel subject to a temperature below Ac1 and then cooling it at any desired rate to increase softness and ductility by reducing the brittleness of the martensite." Tempering allows some of the carbon atoms in the strained martensitic structure to diffuse and form iron carbides or cementite. This reduces the hardness, tensile strength, yield strength, and stress level but increases the ductility and toughness. Tempering temperatures and times are interdependent, but tempering is normally done at temperatures between 175°C and 705°C and for times from 30 minutes to 4 hours, depending on the composition of the steel. Alloying elements generally slow down tempering, which will be discussed in later chapters.

Apart from the regular heat treatments, there are isothermal heat treatments, which produce variety of microstructure and properties. In these treatments, the sample is held above the martensite start temperature and subsequently cooled before it touches the starting curve of softer product transformation or allowed to transform to softer products, which could be done by quenching the steel in a high temperature bath of molten salt or metal. The softer products thus formed have different morphology and characteristics. The IT diagram is useful in helping to predict the resulting microstructures in isothermal heat treatments. There are three types of isothermal heat treatments known as martempering, ausforming, and austempering.

2.6 Martempering

Martempering is not a tempering operation but a treatment that leads to low levels of residual stresses and minimizes distortion and cracking. Martempering is also known as marquenching. The main aim of martempering is primarily to minimize distortion and cracking due to thermal shock loads of heat-treated steels during hardening processes. The steps include the following: (i) austenitize the steel at the appropriate temperature based on iron carbon diagram; (ii) quench to a temperature just above the Ms (usually, into an oil or molten salt bath) based on the IT (TTT) diagram and critical cooling rate; (iii) hold in the quenchant to obtain uniform temperature throughout the cross section; and (iv) cool at a moderate rate through the martensite transformation region in such a way that the transformation of softer products will not start. As the sample is cooled (quenched) from just above Ms temperature and is held for certain period for temperature distribution to become uniform throughout the cross section the distortion and cracking due to thermal shock will be minimum. After martempering, the steel is tempered. Figure 2.8 schematically shows the martempering process.

2.7 Austempering

Austempering is designed to produce bainite in carbon steels. Bainite is similar to pearlite having ferrite and cementite but with different morphology. There are two types of bainite; lower and upper. The hardness and strength of bainite are comparable to hardened and tempered martensite with

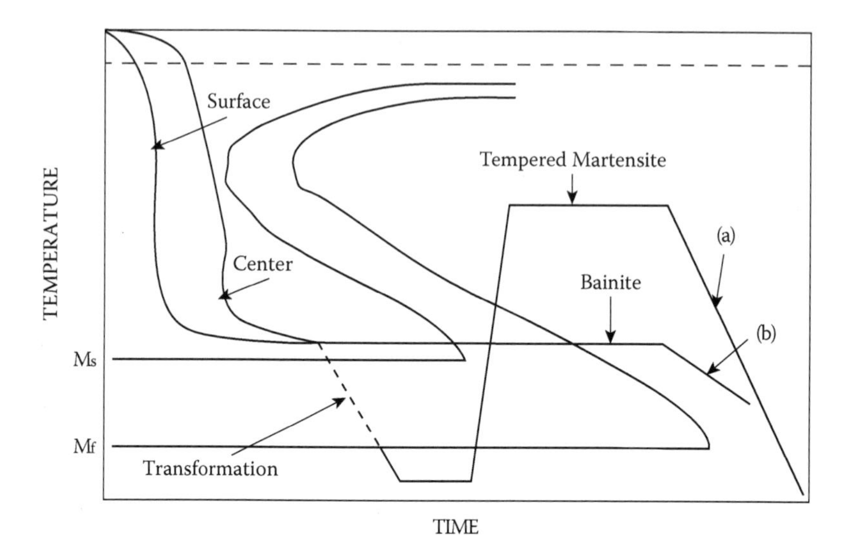

FIGURE 2.8
Schematic representation of (a) martempering process (b) austempering process

better ductility, toughness, and uniform mechanical properties than tempered martensite. In addition, austempered products do not require tempering.

Austempering consists of the following steps: (i) austenitize the steel at the appropriate temperature; (ii) quench to a temperature just above the Ms in a slat bath; (iii) hold isothermally in the salt bath until austenite transforms to bainite; and (iv) quench or air cool to room temperature. Figure 2.8 schematically shows the austempering process.

The essential requirement is that the cooling rate must be fast enough to avoid the formation of pearlite and the isothermal treatment must be long enough to complete the transformation to bainite.

2.8 Ausforming

Ausforming, also known as low-temperature thermomechanical treatment (LTMT), involves the deformation of austenite in the metastable bay between the ferrite and bainite curves of the IT (Figure 2.9) diagram. After deformation, the steel is cooled rapidly to form martensite. The combined effect of austenite deformation and high dislocation density results in a very fine martensite microstructure with high yield strength and high toughness.

The ausforming process cannot be carried out for all types of steel. It can be done only for steels which show distinct ferrite and bainite transformation

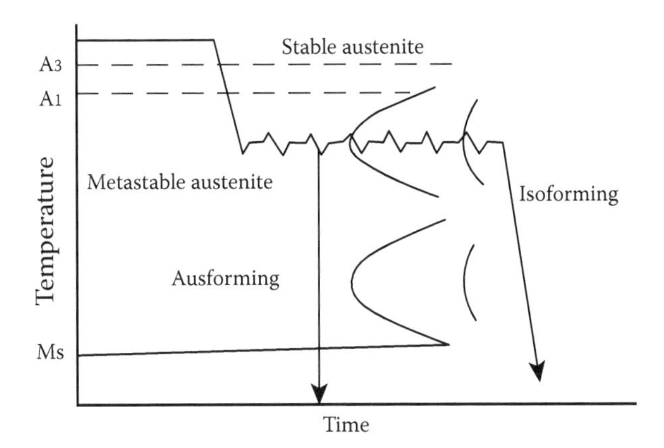

FIGURE 2.9
Schematic representation of ausforming process and isoforming

curves. Alloying elements produce such distinct bays in steels, which will be discussed later in this book.

2.9 Isoforming

This is similar to ausforming, with the deformation of austenite continued at the temperature where austenite transforms to ferrite and iron carbide (Figure 2.9). Martensite is not formed. The microstructure of fine ferrite and spheroidal carbides gives a large toughness improvement.

2.10 Precipitation Heat Treatment

Precipitation heat treatment is less common in carbon steels because the desired precipitates generally are carbides of alloying elements rather than intermetallics. However, some of the low and high alloyed carbon steels containing chromium, molybdenum, niobium, vanadium, and so on respond to precipitation hardening. Figure 2.10 schematically represents the precipitation hardening treatment.

All the alloys will not respond to precipitation hardening also called age hardening. The alloy should satisfy few conditions (i) the system should have negative solid solubility, that is, the solvus curve should have negative slope (Figure 2.10a); (ii) the precipitate formed should be hard intermetallic or carbide; and (iii) the amount of precipitate should be sufficient to be distributed in the parent matrix.

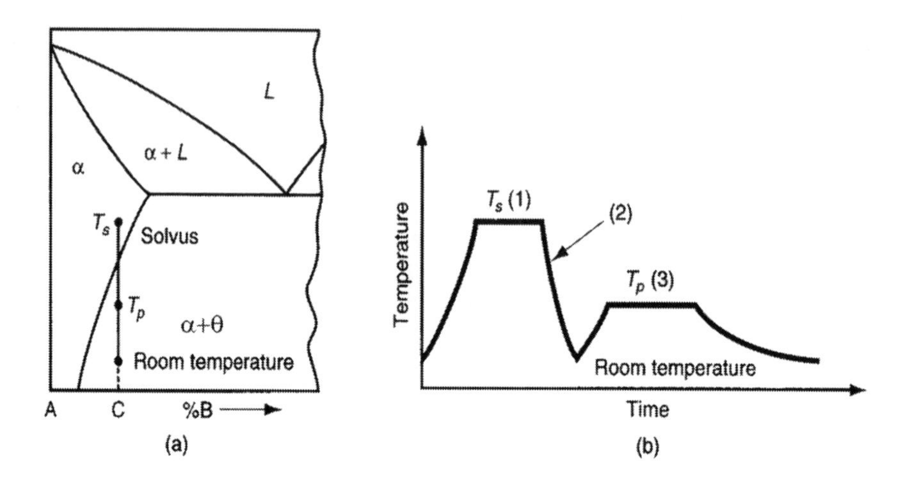

FIGURE 2.10
Schematic representation of (a) phase diagram showing negative solvus line and (b) solutionizing and aging

The alloy with composition C (Figure 2.10a) when cooled under equilibrium conditions has alpha (α) and coarse theta (θ) phase as shown in Figure 2.10. On heating above the solvus line, to a temperature of Ts as shown in Figure 2.10, the second phase theta dissolves in alpha since the solubility limit at the temperature Ts is high. Upon sudden quenching, the second phase theta has no time for precipitation and results in supersaturated alpha (Figure 2.11). Again, on heating to temperature Tp and slow cooling results in finely distributed second phase theta in alpha matrix, as shown in Figure 2.11.

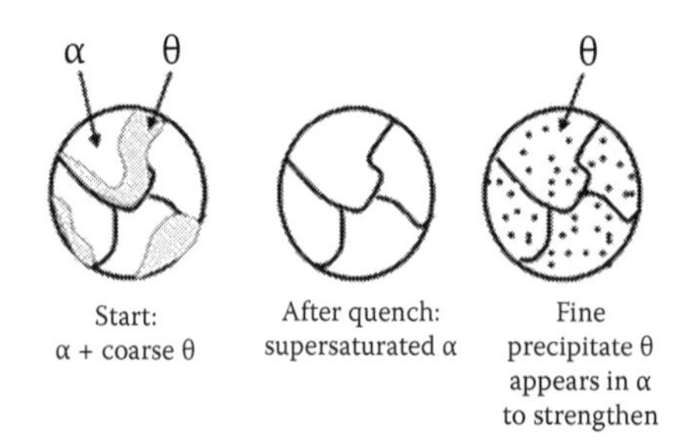

FIGURE 2.11
Microstructural changes during precipitation hardening (schematic)

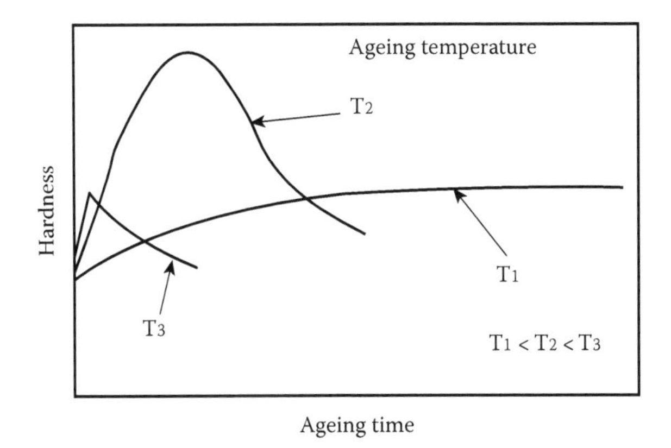

FIGURE 2.12
Schematic representation of aging behavior

The fine hard second phase obstructs the motion of dislocations and increases the strength of the alloy. The mechanism of hardening is beyond the scope of the book. The aging is like tempering of hardened steels depending on temperature and time. Figure 2.12 schematically represents the aging behavior.

3

Surface Hardening

3.0 Introduction

Surface hardening, a heat treatment process that includes a wide variety of techniques, is used to improve the wear resistance of parts without affecting the softer, tough interior of the part. This combination of hard surface and resistance to breakage upon impact is useful for cam or ring gear, bearings or shafts, parts for turbine applications, and automotive components that must have a very hard surface to resist wear, along with a tough interior to resist the impact that occurs during operation. Most surface treatments result in compressive residual stresses at the surface that reduce the probability of crack initiation and help arrest crack propagation at the case-core interface. Further, the surface hardening of steel can have an advantage over through hardening because less expensive low-carbon and medium carbon steels can be surface hardened with minimal problems of distortion and cracking associated with through hardening of thick sections. There are two type of surface hardening: one involves heating the surface alone to the austenitizing temperature and quenching, called "case hardening" and the second involves heating the surface to austenitizing temperature with hardening species such as carbon, nitrogen, boron, and so on, and subsequently quenching based on needs, called "chemical surface hardening."

3.1 Case Hardening

In the case of hardening processes, the composition of the material is unchanged while only the surface is hardened by austenitizing and subsequent quenching. The inner part will not reach the austenitizing temperature and hence even upon quenching, the core will retain the soft and ductile ferrite-pearlite phases as shown in Figure 3.1. The depth (thickness) to which it forms martensite is called "case depth."

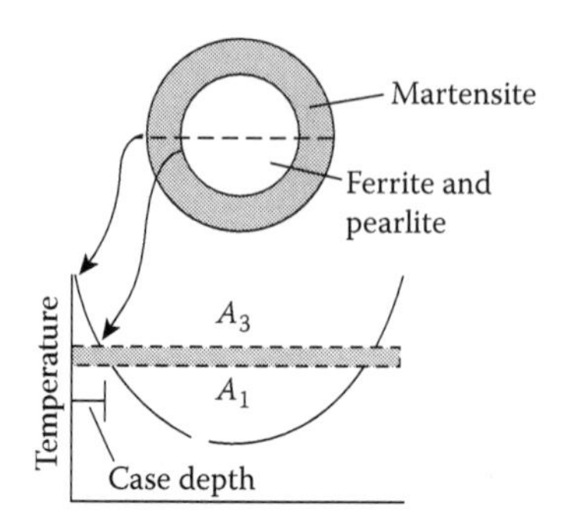

FIGURE 3.1
Schematic representation of case hardened microstructure across its cross section

The selective heating of surface is achieved by induction coils in the case of induction hardening, oxy-acetylene torch in the process of flame hardening as shown in Figure 3.2. Apart from induction coils and oxy acetylene flame, heating by high-frequency resistance, electron beam, and laser beam are also used for heating the surface leaving the core unaffected.

Generally case hardening process is applied to 0.30–0.50% C containing steels to get hardness values in the range 50–60 HRc. Steels with less than 0.3% carbon cannot be case hardened since the critical cooling rate required to achieve martensite is not practically possible. So, for hardening such low carbon steels, chemical surface hardening methods are used.

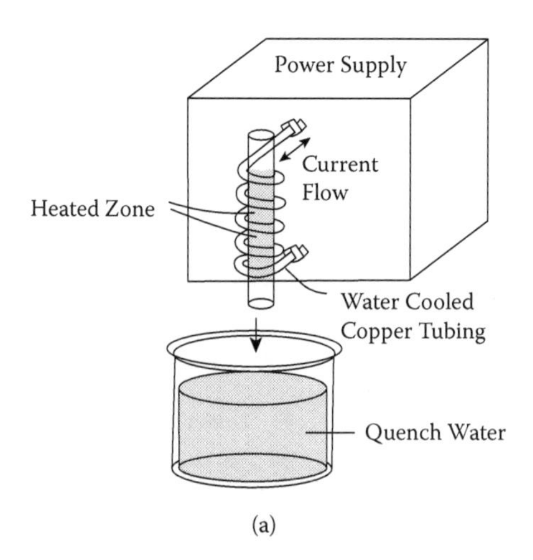

FIGURE 3.2
Schematic representation of (a) induction hardening and (b) flame hardening.

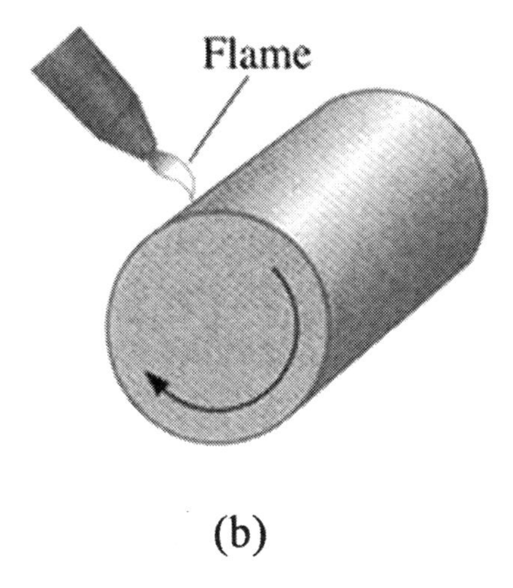

FIGURE 3.2 (Cont.)

3.2 Chemical Surface Hardening

In chemical hardening methods, the chemical composition of the surface is modified by diffusing hardening interstitials such as carbon, nitrogen, boron, and so on. The most widely known chemical surface hardening process is case carburizing practiced since ancient times. Case carburizing involves the diffusion of carbon into the surface layers of a low carbon steel by heating it in contact with a carbonaceous material. Carburizing is carried out at temperatures in the range of 825–925°C in solid, liquid or gaseous media but, in each treatment, the transport of carbon from the carburizing medium takes place via the gaseous state, usually as CO. Carburizing results in high carbon at the surface and low carbon at the core. Following the carburizing operation, the components are subjected to hardening heat treatments resulting in hard martensite at the surface due to high carbon and soft ferrite and pearlite in the core due to low carbon content.

Another most important case hardening process is nitriding. As its name suggests, nitriding involves the introduction of atomic nitrogen onto the surface of a steel but, unlike in carburizing, it is carried out in the ferritic state at temperatures of the order of 500–575°C. Like carburizing, it can be performed in solid, liquid or gaseous media but the most common method involves ammonia gas (gas nitriding), which dissociates to form nitrogen and hydrogen. Nascent, atomic nitrogen diffuses into the steel, forming nitrides in the surface region. Nitriding is carried out on steels containing strong nitride-forming elements such as aluminium, chromium and vanadium which form their respective nitrides in the nitride layer. Given the low temperature involved in the process, nitriding can be carried out on through-hardened steels after the conventional hardening and tempering treatments have been applied.

Another common process is carbonitriding (nitro-carburizing). Carbonitriding can be regarded as a variant of gas carburizing in which both carbon and nitrogen are introduced onto the steel surface. This is achieved by the introduction of ammonia gas into the carburizing atmosphere, which liberates nascent nitrogen. Carbonitriding is carried out at temperatures of the order of 850°C. The introduction of nitrogen produces a significant increase in the hardenability of the case region such that high surface hardness levels can be produced even in steels of relatively low alloy content.

In the same way, if the other interstitial element boron is diffused to the surface of steel, it is called boriding or boronizing. The boron diffused surface contains metal borides, such as iron borides, nickel borides, and cobalt borides. As pure materials, these borides provide extremely high hardness and wear resistance. These favorable properties are manifested even when they are a small fraction of the bulk solid. Boronized steel parts being extremely wear resistant will often last two to five times longer than components treated with conventional heat treatments such as hardening, carburizing, nitriding, nitrocarburizing or induction hardening. Most borided steel surfaces will have an iron boride layer with hardness ranging from 1200–1600 HV.

Apart from these processes, elements such as chromium, (called as chromizing), aluminium, (called as aluminizing), silicon (called as siliconizing), are diffused onto the steel surface for achieving specific properties such as wear, corrosion, and high temperature resistance.

4

Plain Carbon Steels

4.0 Introduction

The term steel is usually taken to mean an iron-based alloy containing carbon in amounts less than about 2%. Plain carbon steels (sometimes also termed carbon steels, ordinary steels, or straight carbon steels) can be defined as steels that contain other than carbon, only residual amounts of elements such as silicon and aluminum added for deoxidation and others such as manganese and cerium added to counteract certain deleterious effects of residual sulfur. However, silicon and manganese can be added in amounts greater than those required strictly to meet these criteria so that arbitrary upper limits for these elements are set. Usually, 0.60% silicon and 1.65% manganese are accepted as limits for plain carbon steel.

Carbon is cheap but an effective hardening element for iron and hence a large tonnage of commercial steels contain very little alloying elements. Figure 4.1 shows the effect of carbon on the strength and ductility.

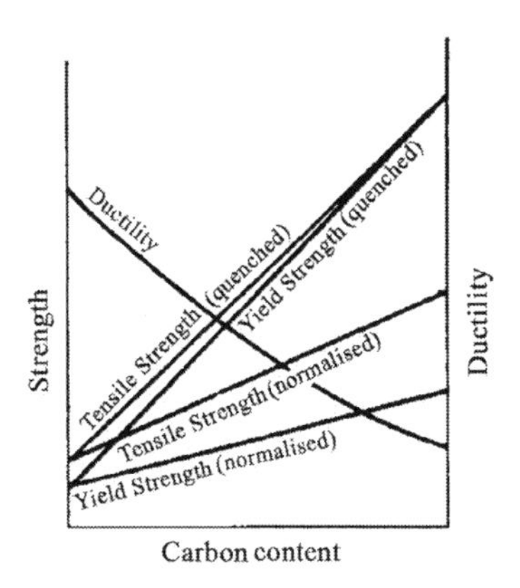

FIGURE 4.1
Effect of carbon on mechanical properties of steel (Schematic)

Plain carbon steels are classified based on their carbon content as low, medium and high: low-carbon steel or mild steel, < 0.3%C, medium-carbon steel, 0.3% ~ 0.6%C, and high-carbon steel, > 0.60% C. Most of the plain carbon steel is used in the hot finished condition, that is, straight from hot rolling without subsequent cold rolling or heat treatment. This is the cheapest form of steel. However, hot finished condition is usually restricted to low carbon and medium carbon grades, because of the loss of ductility and weldability at high carbon contents.

4.1 Low Carbon (or) Mild Steels

These generally contain less than about 0.25% C and are unresponsive to heat treatments intended to form martensite and strengthening is accomplished by cold work. Microstructures consist of ferrite and pearlite constituents. As a consequence, these alloys are relatively soft and weak, but have outstanding ductility and toughness and in addition, they are machinable, weldable, and, of all steels, are the least expensive to produce.

4.2 Medium Carbon Steels

Medium carbon steels (0.25–0.55% carbon) are capable of being quenched to form martensite and subsequently tempered to develop toughness with good strength and hence the medium carbon steels are often used in heat treated (quenched and tempered) conditions. Tempering in higher temperature regions (i.e., 350–550°C) produces a spherodized carbide which toughens the steel sufficiently for specific applications. In addition, ausforming of medium carbon steels produce even higher strengths without significantly reducing the ductility.

The plain medium carbon steels have low hardenability and can be successfully heat treated only in very thin sections and with very rapid quenching rates. Additions of chromium, nickel, and molybdenum improve the capacity of these alloys to be heat treated to a variety of strength–ductility combinations. These heat-treated medium carbon steels are stronger than the low carbon steels, with a sacrifice of ductility and toughness.

4.3 High Carbon Steels

The high carbon steels, normally having carbon contents between 0.60 and 1.4%, are the hardest, strongest, and least ductile of carbon steels. They are almost always used in a hardened and tempered condition and, as such, are especially wear resistant and also capable of holding a sharp cutting edge. The high carbon steels are usually tempered at 250°C to develop considerable strength with sufficient ductility. Their limitations are poor hardenability and rapid softening properties at moderate tempering temperatures and so, the tool and die steels are of high carbon steels, usually containing chromium, vanadium, tungsten, and molybdenum. These alloying elements combine with carbon to form very hard and wear resistant carbide compounds, which are dealt with in Chapter 8.

4.4 Applications as Engineering Material

Typical applications of plain carbon steels, include automobile body components, structural shapes (I-beams, channel and angle iron), and sheets that are used in pipelines, buildings, bridges, and tin cans. They typically have a yield strength of 275 MPa, tensile strengths between 415 and 550 MPa and a ductility of 25%. Low carbon steels are the most important group of hot finished plain carbon steels. Hot rolled low carbon steel sheet is an important product and used extensively for fabrication where surface finish is not of prime importance. Cold rolling is used as final pass where better finish and the additional strength from cold working are needed. However, for high quality sheet to be used in intricate pressing operations, it is necessary to anneal the cold worked steel to cause the ferrite to recrystallize. This is done below the A1 temperature (subcritical annealing). The other important field of application of plain carbon steels is as castings. Low carbon cast steels containing up to 0.25% carbon are widely used as castings for miscellaneous jobbing as reasonable strength and ductility levels are readily obtained. Yield strengths of 240 MPa and elongations of 30% are fairly typical for this type of steel. Of all the different steels, those produced in the greatest quantities fall within the low carbon classification.

Medium carbon steels are used in railway wheels and tracks, shafts, gears, crankshafts, dies for closed die or drop forgings and other machine parts and also for high strength structural components calling for a combination of high strength, wear resistance, and toughness. High carbon steels are utilized as cutting tools and dies for forming and shaping materials, as well as in knives, razors, hacksaw blades, springs, and high strength wire.

Though the plain carbon steels are widely used because of their properties, they have certain lacuna. They are not corrosion resistant, lose their strength upon heating, limited hardenability, which restricts section thickness, and so on. With a view to increase specific properties, alloying elements other than carbon are added which are called as alloy steels. Although it is a misnomer, it is well accepted in the society. Effect of alloying elements on steels are discussed in the next chapter.

5

Effect of Alloying Elements in Steel

5.0 Introduction

The purpose of adding alloying elements in steel is:

1. To increase strength
2. To increase toughness and plasticity
3. To increase hardenability
4. To decrease quench hardening capacity
5. To increase rate of strain hardening
6. To increase machinability
7. To increase abrasion (wear) resistance
8. To decrease warping and cracking
9. To increase magnetic permeability
10. To decrease hysteresis loss
11. To increase corrosion resistance
12. To increase oxidation resistance.

Although all of the properties are listed in general, each alloying element plays its own role in increasing the specific properties for specific requirements.

5.1 Effect of Alloying Elements on Iron Carbon Diagram

It would be impossible to include a detailed survey of the effects of alloying elements on the iron carbon equilibrium diagram in this chapter. Even in the simplest version, this would require analysis of a large number of ternary alloy diagrams over a wide temperature range. However, iron binary equilibrium systems fall into two main categories: open and closed

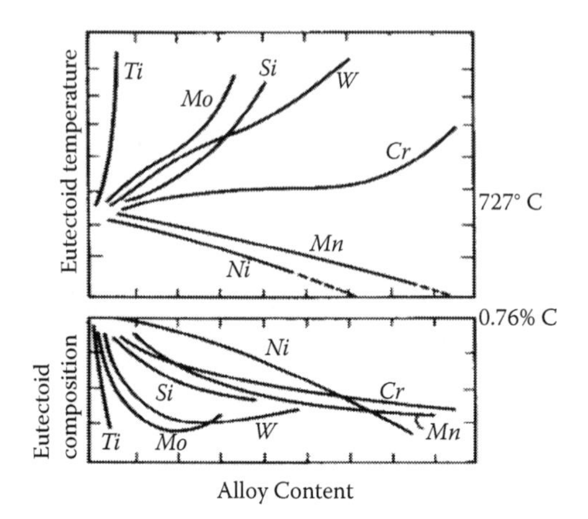

FIGURE 5.1
Effect of alloying elements on eutectoid temperature and composition (schematic)

γ-field systems, and expanded and contracted γ-field systems. This approach indicates that alloying elements can influence the equilibrium diagram in two ways: (a) some elements by contracting the γ-field encourage the formation of ferrite over wider compositional limits and these elements are called α-stabilizers. (b) some other elements by expanding the γ-field encourage the formation of austenite over wider compositional limits and these elements are called γ-stabilizers. Figure 5.1 schematically represents the effect of alloying elements on the steel portion of the iron carbon diagram especially on eutectoid temperature and composition.

It is clear from Figure 5.1, that all alloying elements decrease the eutectoid composition (0.8%C) of the steels. In the same way, except nickel and manganese, all the other alloying elements increase the eutectoid temperature (727°C). Based on this fact, the alloying elements are classified broadly as ferrite stabilizers and austenite stabilizers. The ferrite stabilizers increase the eutectoid temperature while austenite stabilizers decrease it.

5.2 Ferrite Stabilizers

Ferrite stabilizers can be divided into two groups based on their tendency to form carbides. Aluminium and silicon will dissolve in ferrite matrix but do not form any carbides at steel melting temperature, whereas chromium, molybdenum, tungsten, tantalum, vanadium, niobium, zirconium and titanium will dissolve in ferrite matrix and also tend to form carbides. All ferrite stabilizers shrink the austenitic field and make steel non heat treatable (or) make a narrow range of temperature for austenitizing. Chromium around 20%, molybdenum around 7%, tungsten around 12%, silicon around 8.5%, titanium around 1%, niobium around 1%, and vanadium around 4.5% eliminate austenitic field completely and make the steel non-heat-treatable.

Aluminium: Traditionally, aluminium is added as deoxidizer during the melting of steel. Normally it is present as aluminium oxide (alumina) or as aluminium nitride (AlN). These precipitates inhibit austenitic grain boundaries at high temperatures, and by pinning these boundaries they prevent excessive grain growth. On transformation to ferrite and pearlite, grain sizes around 12 ASTM (5–6 μm diameter) can be achieved with as little as 0.03% AlN in steel.

The amount of aluminium rarely exceeds 0.01 to 0.07%, except in specialized steels for nitriding or forging applications. At present there is a great commercial interest in aluminium additions, of the order of 0.5–2%, to low carbon high strength strip steels to produce a highly desirable multiphase microstructure. Though there has been some work on the effect of aluminium in steel microstructures, there has been no reports on addition of aluminium as alloying element in steels. The solubility of aluminium in ferrite is more as compared to austenite and it is a ferrite stabilizer. It will not form any carbides in steels.

Silicon: Like aluminium, silicon is added as deoxidizer. The residue silicon is present as its oxide. Silicon is a ferrite stabilizer and it is one of the ferrite stabilizers that does not form carbides. Silicon is normally added in magnetic steels as it improves magnetic properties and in valve steels as it improves the high temperature properties of steels. Precaution has to be taken while adding silicon because silicon is a powerful graphatizer that accelerates metastable cementite to reach its stable state as iron and carbon (graphite).

Chromium, molybdenum, tungsten, tantalum, vanadium, niobium, zirconium and titanium: All of these elements will dissolve in ferrite in more amounts than in austenite. All these elements have great influence on increasing the eutectoid temperature (Figure 5.1). In the presence of carbon these ferrite stabilizes form their respective carbides, which are more stable than cementite. The carbide forming tendency of these elements are as follows: Mn < Fe < Cr < Mo < W < Ta < V < Nb < Zr < Ti. These carbides improve strength, hardness, wear resistance, and high temperature strength of the steel. Apart from increasing mechanical properties, these stable carbides lock austenite grain boundaries, and thus allow much finer ferrite grain sizes to be achieved when the austenite transforms.

5.3 Austenite Stabilizers

Manganese, nickel, and nitrogen are the most common austenite stabilizers. They reduce the eutectoid temperature and widen the gamma field. They make the decomposition of austenite sluggish and the formation of ferrite and pearlite is avoided during cooling. Solubility of these elements in gamma phase is much higher and less in ferrite

Manganese: Manganese is generally added as deoxidizer during the melting of steels. It is also added to counteract the deleterious effect of sulfur. Manganese about 1% will always be present in plain carbon steels and up to this limit it is not considered as an alloying element. Manganese is an austenite stabilizer though it dissolves in ferrite to a limited extent. Manganese causes the eutectoid composition to occur at lower carbon contents, and so increases the proportion of pearlite in the microstructure. Manganese is also an effective solid solution strengthener, and also has a grain refining influence.

Nickel: Nickel is an austenite stabilizer and about 8% nickel stabilizes austenite at room temperature. Nickel is an important alloying element in austenitic stainless steels and in maraging steels.

Nitrogen: All steels contain nitrogen in traces, and it improves the mechanical and corrosion properties of steel. Nitrogen is being used as an alloying element in steel since 1940s as a substitute for nickel to produce stainless steels. Nitrogen is a very strong austenite former and also significantly increases the mechanical strength. It also increases resistance to localized corrosion, especially in combination with molybdenum. In ferritic stainless steels, nitrogen strongly reduces toughness and corrosion resistance. In martensitic grades, nitrogen increases both hardness and strength but reduces toughness. Nitrogen has efficient solubility in steel as a solid solution or as a chemical compound.

While steel is in molten form, nitrogen is present in solution. However, solidification of steel can result in three nitrogen-related phenomena, namely, formation of blowholes, precipitation of one or more nitride compounds and/or the solidification of nitrogen in interstitial solid solution. The maximum solubility of nitrogen in liquid iron is approximately 450 ppm and less than 10 ppm at ambient temperature. The other elements in liquid iron affect the solubility of nitrogen in the molten steel. More importantly, the presence of dissolved sulfur and oxygen limit the absorption of nitrogen because they are surface-active elements. This fact is exploited during steelmaking to avoid excessive nitrogen pickup, particularly during tapping.

The addition of nitrogen to plain carbon steels results in three different microstructures: (i) for low nitrogen additions, the precipitate formed is Fe_3C and the nitrogen enters into the ferrite matrix as interstitial, (ii) for low carbon additions, the precipitate formed is Fe_4N and the carbon goes into the ferrite matrix as interstitial, and (iii) for elevated concentrations of both carbon and nitrogen (both above approximately 0.4%) the matrix is austenite and ferrite and no precipitates are formed. With increasing nitrogen and/or carbon additions, alloys with each of the three different microstructures, increased hardness, and wear resistance are formed.

However, in carbon or low alloy steels, dissolved nitrogen being undesirable, is kept to a minimum because high nitrogen content may result in inconsistent mechanical properties in hot rolled products, embrittlement of the heat affected zone (HAZ) of welded steels, and poor cold formability. In particular, nitrogen can result in strain ageing and reduced ductility of cold rolled and annealed steels although it improves yield, grain size, and mechanical properties of steels.

In general limited amounts of nitrogen if present as fine and coherent nitrides or carbonitrides of iron or alloying elements have a beneficial effect on the mechanical properties. Nitrides of aluminum, vanadium, niobium and titanium result in the formation of fine grained ferrite which improves mechanical properties, lowers the ductile to brittle transition temperature and improves toughness. But if nitrogen is present in ferritic matrix as interstitial element, the ductile to brittle transition temperature increases and toughness decreases. Therefore, it is necessary to carefully control, not only the nitrogen content, but also the form in which it exists, in order to optimize impact properties.

When nitrogen is added to austenitic steels, it can simultaneously improve fatigue life, strength, work hardening rate, wear and localized corrosion resistance. High nitrogen martensitic stainless steel shows improved resistance to localized corrosion, like pitting, crevice and intergranular corrosions, over their carbon containing counterparts. Because of this, the high nitrogen steels are considered as a new promising class of engineering materials.

The effect of nitrogen in steel can be either detrimental or beneficial, depending on the presence of other alloying elements in it, the form and quantity of nitrogen present, and the required behavior of the particular steel product. When properly employed, nitrogen is a friendly addition to steels.

Apart from these elements, inert metals ruthenium, rhodium, palladium, osmium, iridium and platinum stabilize the austenite but they are too expensive and hence not used in steels as alloying elements.

5.4 Neutral Stabilizer

Cobalt can dissolve both in ferrite (about 80%) and in austenite (100%). It stabilizes austenite when the amount exceeds 90%. It has no role to play in increasing the eutectoid composition or temperature. Like aluminium and silicon, it also does not form any carbides. Hence it can be classified as a neutral stabilizer.

Cobalt is the only alloying element that increases the critical cooling rate of steel and accelerates pearlitic transformation thus reducing hardenability. Also, it has a tendency to graphitization and is a very expensive component, hence not used as an alloying addition in normal steels. It is never used in the standard heat treatable steels. Hence cobalt is not a popular element that is commonly added to steels. Whatever way that it finds use in tool steels, 18% nickel maraging steels, and several other ultrahigh strength steels, its adverse effect on hardenability overcome by the remaining alloying constituents.

The presence of cobalt in steel improves its durability and hardness at higher temperatures, reduces the fall in hardness of both austenite and ferrite due to increase in temperature, and thus it increases hot hardness. Therefore, cobalt is used as a supplement to some grades of high speed steels and tool steels, so that tools made from cobalt bearing steel can operate at high temperatures, maintaining their cutting edge. Cobalt is also a component of creep resistant steels. Magnetic steels containing from 9% to 40% cobalt have been used for compass needles, hysteresis motors and electrical instrumentation. Cobalt is used in stainless steel of grade 348, which contains 0.20% of cobalt. The dissolved cobalt in ferritic or austenitic matrix has a high-work hardening sensitivity, combines with the carbide fraction and allows to achieve excellent wear resistance along with a high degree of corrosion resistance.

5.5 Effect of Various Elements as Dissolved in Ferrite Matrix

In the absence of carbon, all the alloying elements irrespective of whether ferrite or austenite stabilizer increase the strength of the matrix by solid solution strengthening as shown in Figure 5.2.

The increase in strength/hardness is only due to solid solution hardening. Phosphorous is the most effective solid solution hardener but above 0.25% it forms Fe_3P, which is extremely brittle and reduces the ductility considerably. Therefore, it will be normally restricted to below 0.25%. Solid solution strengthening effect increases tremendously if little carbon is present.

5.6 Effect of Various Elements as Carbides in Ferrite/Austenite Matrix

All carbides are brittle in nature and alloy carbides have less tendency of coalescence and also stronger than cementite (Fe_3C). The dissolution of these carbides in austenite is sluggish and needs high temperature for austenitizing, which is the main reason for increase in eutectoid temperature. The type of carbides are MC, M_7C_3, M_6C, M_3C, and $M_{23}C_6$, where M refers to metal. The type of carbides formed is based on the

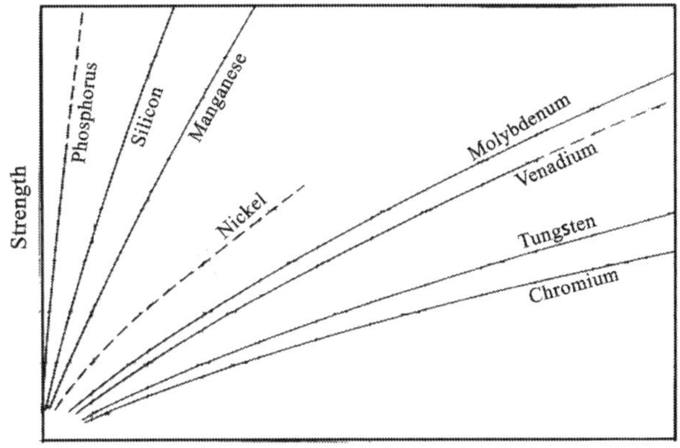

FIGURE 5.2
Effect of alloying elements on solid solution strengthening of the steel (schematic)

amount of the element and the carbon content. The non-carbide-forming elements such as nickel, silicon, and aluminium accelerate graphatization of cementite.

Chromium carbides: Chromium forms $(FeCr)_3C$, which is orthorhombic having 15% Cr, $(FeCr)_7C_3$, which is trigonal having 36% Cr, and $(FeCr)_{23}C_6$, which is cubic having 70% Cr. The chromium carbides tend to coalesce easily and form coarse carbides, which depends on cooling rate and so the strengthening due to chromium carbides is controlled by the mode of cooling. Air cooled chromium steels give higher hardness than furnace cooled because air cooled steels have fine carbides and higher amount of pearlite as compared to furnace cooled which have coarse carbides. The chromium forms an intermetallic phase (σ) with iron and chromium content varying between 20 and 80% in the σ phase. Usually, this intermetallic phase is observed in stainless steels.

Molybdenum carbides: Molybdenum forms carbides such as $(FeMo)_6C$, or $Fe_7Mo_5C_2$. Molybdenum forms Fe_3Mo_2 intermetallic, which is able to dissolve 0.1% carbon and this intermetallic is expected to precipitate in low carbon age hardenable molybdenum containing steels.

Tungsten carbides: Like molybdenum, tungsten forms $(FeW)_6C$ or $Fe_7W_5C_2$ and is primarily found in High Speed Steels (HSS). In the absence of chromium and vanadium or even if present and are dissolved in carbide replacing iron, when tungsten is about 18% it forms $(FeW)_3C$.

Tantalum carbides: Tantalum forms TaC in steels, which plays an important role in lowering DBTT through its effect on prior austenitic grain size refinement and also improves the strength and toughness. However, higher tantalum content can result in the formation of extremely coarse TaC during the melting process, which is deleterious to the mechanical properties. With the increase of tantalum content, the content of chromium rich $M_{23}C_6$ carbides decreases. Because of its high cost tantalum is not used in low alloy steels. However of late, tantalum has been increasingly used as an alloying element only in special steels – super strong, corrosion resistant, and refractory. The addition of tantalum increases the creep resistance of steels.

Vanadium carbides: The iron, vanadium, and carbon diagram shows the absence of intermetallic compounds of iron and vanadium unlike in chromium, tungsten, and molybdenum systems. Vanadium addition introduces $(VFe)_4C_3$. Vanadium is by far the most used element to improve wear resistance in tool steels. At present there are several vanadium-alloyed tool steels with up to 18% vanadium in the market.

Niobium carbides: The effect of niobium on steel is similar to that of tantalum. Niobium is a strong ferrite and carbide former like tantalum and even a small addition of niobium to steels is effective for austenitic grain size refining due to the precipitation of δ-niobium-carbide and half the amount (in wt%) of niobium is sufficient compared to tantalum, to obtain the same fine grain size. Niobium, about 0.03 to 0.04%, is added to refine remarkably the cast structure and austenite structure of steel.

Addition of niobium (up to 1%) to chromium steel and in stainless steels improves its ductility and corrosion resistance by preventing precipitation of chromium carbides along grain boundaries. An addition of 0.7% niobium decreases the ductile to brittle transformation temperature of steel to 80°C below zero by refining the grain size of the matrix. During welding of austenitic stainless steels, niobium stabilization is better than titanium stabilization as titanium can be easily oxidized during welding.

Niobium addition plays major role on microstructures and formability of TRIP and HSLA steels. However, higher contents of more than about 2.5% of niobium cause considerable difficulties during melting. Niobium carbides directly precipitate at higher temperatures before the ferritic crystallization of the melt starts and the melt becomes more and more viscous and pulpy. Due to the viscosity, melts with more than about 3% niobium cannot be poured in moulds. For this reason the niobium content in tool steels is limited to about 2%. There are many attempts to improve the poor solubility of niobium in the melt and until now no experiment has been successful and satisfactory.

Zirconium carbides: Zirconium is being used as an alloying element in steels since the early 1920s, but has never been universally employed, unlike chromium, niobium, titanium, and vanadium. Zirconium is highly reactive and has a strong affinity, in decreasing order, for oxygen, nitrogen, sulfur, and carbon. Zirconium is a strong carbide former and forms zirconium carbide (ZrC), containing carbon in the range of 6 to 12%. Zirconium is used mainly in micro alloyed steels. Zirconium also forms zirconium nitride (ZrN) when nitrogen content is above 9%. There is extensive solid solubility between ZrC and ZrN to form Zr (CN) precipitates. Zirconium carbide, nitride and carbo nitride inhibit grain growth and prevent strain aging. However, their use for either of these reasons is limited. Because of its relatively high price and also due to the availability of cheaper replacements, general acceptance of zirconium for use as an alloying element in steels is limited. However, recent days there has been a renewed interest in the addition of zirconium to steels especially micro alloyed steels because of the following facts:

Carbides, nitrides, and carbonitrides of zirconium effectively pin the austenite grains and avoid grain coarsening during heat treatment or hot working process. Addition of zirconium to plain carbon steels effectively refines the grain as compared to other alloying elements and also reduces the interstitial carbon and nitrogen from ferrite. These two effects reduce the ductile to brittle transformation temperatures to a greater extent in steels containing zirconium.

Titanium Carbides: Titanium is the strongest ferrite and carbide/nitride former. It forms more stable TiC, TiN, and Ti (CN). There are no complex carbides of iron and titanium. Titanium bearing steels contain less cementite (less pearlite) because titanium combining with carbon or nitrogen forms the highly stable carbide (TiC) or nitride (TiN). Carbon scavenged in this manner is about 0.25% of the available titanium. This means that pearlite is completely absent in steels containing 0.25% carbon if titanium amount is equal to more than four times (1%) the carbon content.

Titanium is also used for the purpose of grain refining in many steels especially in micro alloyed steels. Titanium increases the grain coarsening temperature. In this respect it is much more effective than aluminium when concentrations of either element exceeds 0.035%.

Titanium as an alloying element is mainly used in stainless steel for carbide stabilization. In austenitic stainless steels, TiC, which is quite stable and hard to dissolve in steel, tends to minimize the occurrence of intergranular corrosion when approximately 0.25%–0.60% Ti is added. The carbon combines with the titanium in preference to chromium, preventing a tie up of corrosion resisting chromium as intergranular carbides and the accompanying loss of corrosion resistance at the grain boundaries. In addition, it also increases the mechanical properties at high temperatures.

In ferritic grades, titanium is added to improve toughness, formability, and corrosion resistance. In martensitic steels, titanium lowers the martensite hardness by combining with carbon and increases tempering resistance. In precipitation hardening steels, titanium is used to form the intermetallic compounds to increase the strength.

5.7 Summary of Carbides

As a summary, it can be stated that the alloying elements in steels form MC, M_6C, M_7C_3, and $M_{23}C_6$ type of carbides where M represents the metallic element(s). Elements like titanium, niobium, tantalum, and zirconium form MC type of carbides called primary carbides. Less reactive metals like molybdenum, tungsten, and vanadium can also substitute in these primary carbides but weaken the binding forces and degradation reaction to form $M_{23}C_6$ and M_6C type of carbides. Primary carbides with titanium, niobium, tantalum, and zirconium counteract the degradation of MC carbides even at temperatures of the order 1200–1260°C.

$M_{23}C_6$ carbides tend to form with moderate to high chromium content. They form during lower temperature heat treatment and service (760–980°C) leading to degeneration of MC carbides and form soluble carbon residual in the alloy matrix. Preferred sites for formation of these carbides are at grain boundaries and occasionally along twin lines, stacking faults and at twin ends. $M_{23}C_6$ has a complex cubic structure and if carbon atoms are removed, the structure closely resembles topographically close packed (TCP) sigma phase which is deleterious phase in stainless steels. In fact, coherency between $M_{23}C_6$ and sigma is very high and sigma phase very often nucleates on $M_{23}C_6$ particles. Molybdenum and tungsten can substitute to certain extent and is found to have $Cr_{21}(Mo,W)_2C_6$ composition. It has also been shown that considerable quantity of nickel and small portions of cobalt or iron could also substitute for chromium. Formation of $M_{23}C_6$ type carbides at grain boundaries reduces grain boundary sliding and hence improves rupture strength but cellular structure (network) of $M_{23}C_6$ initiates premature rupture failures. The grain boundary $M_{23}C_6$ type carbides are prone to intergranular corrosion.

M_6C carbides have a complex cubic structure and form at comparatively higher temperatures (815–980°C) when molybdenum and/or tungsten contents are high (above 6–8%). M_6C carbides also form when molybdenum or tungsten replace chromium in other carbides. Because M_6C carbides are stable at higher temperatures than $M_{23}C_6$, it is more beneficial to control the grain size during the processing of wrought alloys. M_6C also tends to form $M_{23}C_6$ in some alloys.

The type of carbide formed also depends on the amount of carbon available. The different type of carbides formed with respect to carbon content and temperature.

The properties of the steel entirely depends on the type of carbides and nitrides formed. It should be noted that the formed carbides must be fine and homogeneously distributed in the matrix. Any coarse carbides/nitrides distributed homogenously will lower the properties.

5.8 Dispersed Metallic Particles

Apart from the elements discussed in this chapter, a few elements are added to steels to improve specific properties. These elements will not form any compounds or dissolve in the matrix. They will be present in the steel as metallic particles.

Copper: Copper is soluble in austenite up to 2% and has no solubility in ferrite. At 600°C, the solubility is less than 0.3% and hence rejected as elemental copper. It is neither a ferrite nor austenite stabilizer. It does not form carbides or compounds. It is used as a precipitation hardener (1.5–1.75%) in precipitation hardenable stainless steels.

Lead: Like copper, lead also neither dissolves in ferrite nor forms any compound. About 0.25% lead is added to improve machinability because lead has low cohesive strength and thus easy to remove as chips. However, the addition of lead affects hot working characteristics being a low melting point metal.

5.9 Nonmetallic Inclusions

During steel melting, some elements that are not added intentionally as alloying elements enter the ingots as nonmetallic inclusions. However, a few elements such as manganese, tantalum, zirconium, and titanium are intentionally added to form compounds to improve specific properties.

Manganese: As discussed previously, manganese is added to counteract the effects of sulfur in steels. It forms manganese sulphide (MnS) as elongated inclusions, which act as a miniature notch to aid breaking of chips easily by shear improving machinability. But MnS has deleterious influence on the impact toughness of wrought and welded steels.

Tantalum: Tantalum addition improves pitting resistance in duplex stainless steels by controlling inclusions such as manganese sulphide (MnS), $MnCr_2O_4$, and (Ti,Ca)-oxides known to act as pitting initiators in environments containing chloride ions. The addition of tantalum leads to the formation of (Ta,Mn)-oxysulphides and these inclusions are electrochemically stable.

Zirconium: The main use of zirconium additions to steel is to preferentially combine with sulfur, to avoid the formation of manganese sulphide (MnS) known to have a deleterious influence on the impact toughness of wrought and welded steel. Also, it controls the other nonmetallic oxysulphide inclusions and for the fixation of nitrogen mainly in boron steels.

The function of zirconium is not to remain in solution in steel but to scavenge impurities (oxygen, sulfur, nitrogen) or modify inclusions through the formation of complex sulphides and oxysulphides. The efficacy of zirconium additions will therefore be measured not by the amount of residual zirconium, but by the extent to which inclusions are beneficially modified.

Titanium: Like zirconium, titanium is also used for fixing sulfur in order to reduce its harmful effects. Titanium forms stable compounds with oxygen and sulfur at the steelmaking temperatures. With respect to fixing oxygen, titanium functions similar to aluminium. However, titanium is more expensive and hence rarely used as a deoxidizer.

Aluminium: Aluminium oxide formed due to excess aluminium available after deoxidation pins the austenite grains restricting their growth.

Graphite: Graphite is an inclusion in steels but not in cast irons. Cementite in the pearlite forms as graphite when excess amount of graphatizer such as silicon or nickel are available and if exposed to high temperature for long periods.

5.10 Intermetallic Compounds

Normally, no intermetallic compound is found in plain carbon steels. Nitrides of aluminium, zirconium, vanadium, and titanium form as intermetallics in plain carbon steels mainly with a view to obtain fine-grained structures for some special applications. A few high alloy steels such as maraging steels and precipitation hardening steels have intermetallics of nickel, titanium, and aluminium that help in age hardening of those alloys; this will be discussed in Chapter 8. Some intermetallics like sigma phase, which usually form in stainless steels, are undesirable.

5.11 Alloy Distribution in Austenite

The elements such as nickel, manganese and nitrogen that dissolve in austenite matrix minimize eutectoid temperature and try to stabilize austenite at room temperature. In steels, about 8% nickel retains austenite at room temperature. The ferrite stabilizers forming carbides, if they are undissolved in the austenite, pin grain growth of austenite resulting in fine ferrite and pearlite grains upon cooling. The undissolved nonmetallic inclusions (such as alumina, titanium dioxide, and vanadium oxide) have the same effect as carbides if they are fine and uniformly distributed.

Table 5.1 gives the abstract of the effect of alloying elements on steels.

5.12 Effect of Alloying Elements on Hardening

Most alloying elements forming solid solution in austenite lower the Ms temperature, with the exception of cobalt and aluminium. However, the interstitial solutes carbon and nitrogen have a much larger effect than the metallic solutes. About 1% of carbon lowers the Mf by over 300°C.

The relative effect of other alloying elements is indicated in the following empirical relationship due to Andrews (concentrations in wt%):

TABLE 5.1

Effect of alloying elements on steel

Element	Maximum solubility (in wt%) in pure austenite	Maximum solubility (in wt%) in pure ferrite	Carbides formed	Nonmetallic inclusion	Special Intermetallic Compounds	In elemental state
Aluminium	0.6	30	No	Al_2O_3	AlN	–
Chromium	12–20	infinite	$Cr_{23}C_6$, Cr_7C_3 and Cr_3C	No	Nil	–
Cobalt	infinite	80	No	No	Nil	–
Copper	2	Nil	No	No	Nil	Cu
Lead	Nil	Nil	No	No	No	Pb
Manganese	infinite	10	No	MnS, MnFeO, $MnO.SiO_2$	Nil	–
Molybdenum	3	32	Mo_6C	No	Fe_3Mo_2	–
Nickel	infinite	25–30	No	No	Nil	–
Niobium	~ 0.9	Nil	NbC	No	Nil	–
Nitrogen	~ 200 ppm	less than 10 ppm	No	No	nitrides with Al, Ti, V, Zr	–
Phosphorous	0.5	2.5	No	No	Fe_3P	–
Silicon	2	18.5	No	SiO_2	Nil	–
Titanium	0.76	6	TiC	TiO_2	TiN, Ti(CN)	–
Tungsten	6	32	W_6C, W_3C	No	Nil	–
Tantalum	~ 0.9	Nil	TaC	(Ta,Mn)-oxysulphides	Nil	–
Vanadium	1–2	30	V_4C_3	VO_2	VN	–
Zirconium	~ 4	Nil	ZrC	ZrS	ZrN, Zr (CN)	–

Ms(°C) = 539–423 (%C) – 30.4 (%Mn) – 17.7 (%Ni) – 12.1(%Cr) – 7.5 (%Mo).

The equation applies to a limited class of steels. A better approach is to express Ms in terms of the driving force for transformation.

Above 0.7% C, the Mf temperature is below room temperature and consequently higher carbon steels quenched into water will normally contain substantial amounts of retained austenite. These steels have to be cooled below Mf temperature immediately without time delay to convert the retained austenite to martensite and this is known as subzero cooling.

Time delay in subzero treatment does not result in as complete a transformation to martensite, which is called as stabilization. The retained austenite upon tempering will remain as either austenite or transforms to softer products depending on the alloying elements present. In some cases, the retained austenite may convert to martensite on sudden or impact loading. This will increase the toughness of the steel. This phenomenon is used in some special steels for some particular applications.

When the cooling of a steel is arrested in the Ms–Mf range, the degree of stabilization increases to a maximum with time, and as the temperature approaches Mf, the extent of stabilization increases. However stabilization occurs only when a small amount of martensite is present in the matrix. This is because the formation of martensite plates leads to accommodating plastic deformation in the surrounding matrix, resulting in high concentrations of dislocations in the austenite. Time delay at an intermediate temperature gives time for plastic relaxation, that is, movement of dislocations, as well as the locking of interfacial dislocations by carbon atoms. Interaction of some of these dislocations with the glissile dislocations in the martensite plate boundary makes martensite plate immobile so that the plate cannot grow further.

5.13 Effect of Alloying Elements on Hardenability

All alloying elements increase hardenability except cobalt. There is no clear mechanism why cobalt decreases the hardenability, probably because cobalt addition increases the free energy difference between austenite and ferrite and hence the nucleation kinetics is fast enough to form the softer products. Steel with high alloying elements can form martensite even if they are held isothermally. Such type of martensite is called as isothermal martensite witnessed in maraging steels. The effect of alloying element on hardenability is shown in Chapter 2 (see Figure 2.7). For the same amount of carbon, addition of alloying element has increased the hardenability. But it should be noted that alloying elements increase hardenability but not the hardness. Only carbon has greater influence of increasing hardness of the martensite. Carbon is the only element that increases both hardness and hardenability.

There is a complex effect of carbide formers on hardenability. The carbide formers increase hardenability if they are present in solid solution of austenite or the carbides should dissolve completely in austenite or else hardenability decreases with the increase in concentration of the carbide formers, unless austenitizing temperatures are raised. This is due to the formation of respective carbides that lowers the carbon concentration in austenite. Further, undissolved carbides refine the grain size, and this decreases hardenability all the more.

5.13.1 Carbon Equivalence

Carbon is the most important contributor to hardenability, hardness and strength of steels. Even when other alloying elements are not present, high carbon content can result in high hardness and hardenability. However, other alloying elements also contribute to the overall hardenability of the steel. This effect can be generally quantified by the determination of the carbon equivalence (CE) of the steel. CE is defined by several formulae, and it is important that close attention be paid to the formula being used. The following formula is used in most ASME applications:

$$CE = C + (Mn + Si)/6 + (Cr + Mo + V)/5 + (Ni + Cu)/15$$

It is important that any CE determination is calculated using the actual chemical analysis rather than the maximums specified in materials specifications. If this is not done, the calculation will result in an unrealistically high CE.

5.13.2 Factors Affecting Hardenability

Apart from the alloying elements, there are three factors that affect the hardenability; (i) grain size, (ii) undissolved inclusions and carbides and (iii) inhomogeneity of austenite.

Coarse austenite grains result in better hardenability. In the fine-grained austenite, the grain boundaries can act as the nucleation site for the transformation of softer products. In similar way the undissolved inclusions and/or the carbides also act as nucleation site for the formation of softer products. The austenitizing temperature has to be very high to dissolve the carbides especially of vanadium, niobium and titanium. If the austenitizing temperature is raised with a view to dissolve the carbides, grain growth will result. Although the dissolution of carbides and coarse grains is favorable for better hardenability, the martensite formed will be coarse. The desired properties may not be achieved. Therefore, a balance between these two has to be achieved.

The third factor is the inhomogeneity in austenite. On heating to the austenitizing temperature, the austenite formed may not have the same amount of carbon. The austenite grain having lower carbon content may transform to softer product and ends up with ferrite martensitic structure. The steel sample should be held at the austenitizing temperature to homogenize the austenite. The holding time is very critical in steels with high amounts of tungsten, vanadium or niobium. Longer holding time to achieve homogeneous austenite may decrease the hardness and hardenability. Initially, carbon diffuses from carbides and forms austenite at moderate velocities, then the alloying elements dissolve in ferrite and try to stabilize the ferrite ending up with ferrite and pockets of austenite and after hardening result in ferrite and martensite having lower hardness.

5.14 Effect of Alloying Elements on Tempering

In plain carbon steels, as the tempering temperature increases, the hardness decreases but in alloy steels with carbides (especially chromium, molybdenum), the hardness decreases up to certain temperature and some sudden increase in hardness is observed which is called secondary hardening. Between 200 and 300°C, the rejected carbon from martensite forms $Fe_{2.4}C$ (epsilon carbide), which grows up to 400°C to a length of 300A°. Above 400°C, the epsilon carbide dissolves and forms new alloy carbides of 50A° in size, which increases the hardness. With further increase in temperature, the alloy carbides grow larger and hardness starts decreasing again.

These alloying elements are added to the steel to increase specific properties of steels. Based on the alloy content, the alloy steels are classified as low and high alloy steel. There is no strict rule to classify the low and high alloy steels.

6

Low Alloy Steels

6.0 Introduction

Although it is a misnomer, alloy steels contain significant amounts of alloying elements. The alloy steels are generally classified as low alloy steels and high alloy steels. In general if the amount of total alloying elements exceeds 5% it is called as high alloy steels and below 5% it is referred to as low alloy steels. In this chapter, only low alloy steels used for common applications are discussed. The Society of Automotive Engineers (SAE), the American Iron and Steel Institute (AISI), and the American Society for Testing and Materials (ASTM) are responsible for the classification and specification of steels as well as other alloys. The AISI/SAE designation for these steels consists of a four-digit number, the first two digits indicating the alloy content, while the last two indicate the carbon concentration.

For plain carbon steels, the first two digits are 1 and 0 while alloy steels are designated by other initial two-digit combinations (e.g., 13, 41, 43). The third and fourth digits represent the weight percent carbon multiplied by 100. For example, 1060 steel is a plain carbon steel containing 0.60% C. Various low alloy steels designated as per AISI classification are given below:

Carbon steels 1XXX

Plain carbon steels 10XX

Free machining steels 11XX

Resulfurized and rephosphorized Plain carbon steels 12XX

Manganese Steels 13XX

Nickel steels 2XXX

3.5% Ni 23XX

5% Ni 25XX

Ni-Cr steels 3XXX

1.25% Ni- 0.65% Cr 31XX

3.5% Ni- 1.6% Cr 33XX
Mo steel 4XXX
C- 0.25% Mo 40XX
Cr-Mo 41XX
Ni-Cr-Mo 43XX
C-0.4%Mo 44XX
C-0.55%Mo 45XX
1.8% Ni-Mo 46XX
1.05% Ni-Cr-Mo 47XX
3.5% Ni-Mo 48XX
Cr steels 5XXX
Low Cr (0.35%) 50XX
Medium Cr (0.9% Cr) 51XX
High Cr (1.45% Cr) 52XX
Cr (0.75%) – V 61XX
W steel 7XXX
Triple alloy steels
0.3% Ni-0.4% Cr-0.12% Mo 81XX
0.55% Ni-0.5% Cr- 0.2 Mo 86XX
0.55% Ni-0.5% Cr – 0.25% Mo 87XX
0.55% Ni -0.5%Cr – 0.35% Mo 88XX
3.25% Ni-1.2% Cr- 0.12% Mo 93XX
1% Ni- 0.8% Cr- 0.25% Mo 98XX
Boron Steels (0.0005% B – 5 ppm)
C TS14BXX
0.5% Cr 50BXX
0.8% Cr 51BXX
0.3% Ni-0.45% Cr-0.12% Mo 81BXX
0.55% Ni-0.5% Cr-0.2% Mo 86BXX
0.45% Ni-0.4% Cr-0.12% Mo 94BXX

6.1 Plain Carbon Steels (1XXX Series)

The plain carbon steels are classified as low carbon steels in which carbon% is less than 0.25%; in medium carbon steel carbon content varies between

0.25 and 0.45%; in high carbon steel the carbon content varies between 0.45 and 0.6%; eutectoid steels have 0.8% carbon and hypereutectoid steels have more than 0.8% carbon. In the plain carbon steels, as the carbon content increases, the strength increases but the ductility and the fabrication properties such as weldability, formability decrease.

The low carbon steels have reasonable strength but cannot be hardened by heat treatment. The low carbon steel is widely used as constructional material and is the only alloy used in hot finished condition. The medium and high carbon steels are hardenable and have better strength than low carbon steels.

In 1XXX series there is one classification 11XX, which is called as free machining steels. These steels have slightly higher amounts of sulfur, which increase the machinability of steels by forming elongated manganese sulphide (MnS) inclusions. However, these types of steels have problems during welding and hot forming. These types of steels are used where extensive machining is required.

The other classification in 1XXX series is 13XX called as manganese steels. Generally, in all steels about 1% manganese is always present to counteract the ill effect of sulfur. The sulfur forms iron sulfide (FeS) – a low melting eutectic – which affects the hot working properties and weldability by cracking in hot condition (hot cracking). Addition of manganese forms manganese sulphide (MnS) in preference to FeS. MnS will not melt during hot working and hence does not affect the hot working and welding properties. Moreover, MnS has higher surface tension and higher melting point than FeS and thus improves the weldability by forming isolated pockets of liquid during welding instead of continuous liquid film as in the case of FeS. It is well established that if the surface tension between solid grains and grain boundary liquid is low it forms continuous liquid film and leads to high crack sensitivity, whereas if the surface tension is high, only isolated pockets of liquid forms, which reduces the sensitivity to hot cracking. About 1.6–1.9% manganese is used in low alloy steels to increase strength and ductility, whereas above 2% imparts brittleness.

6.2 Nickel Steels (2XXX)

Nickel is the first element alloyed with iron, which is now removed from the series of low alloy steels due to the high cost of nickel. However, for few special and critical applications this series of alloys are used. Although nickel is an austenite stabilizer, it strengthens ferrite by solid solution strengthening. Addition of nickel increases strength without much decrease in ductility. It also increases impact strength and toughness as well as improves toughness at low temperatures when added in small amounts. And so, Nickel steel has good low temperature strength and high Notch Tensile Strength (NTS). Nickel addition improves steel's resistance to oxidation, corrosion and abrasive resistance. Nickel is heat-resistant, and when alloyed with steel, it increases the heat resistance of the steel. Because nickel reduces eutectoid temperature (approximately 100°C for 1% of Ni), low temperature quenching is sufficient to get full austenite during heat treatment and hence distortion and thermal stresses are less in the components after forming martensite on quenching.

Nickel is a weak hardenability agent and among the standard alloy steels, those containing nickel alone as the principal alloying constituent are rare. Instead, nickel is used in combination with other alloying elements such as chromium, molybdenum, or vanadium to produce steels with excellent combinations of strength and toughness in the quenched and tempered condition. Most of these steels contain around 0.5% nickel, although over 3% nickel is found in some grades. The alloy 2317 can be case carburized to have strong surface and tough core and is used for high strength bolts and high tensile bolts. Alloy 2515 is used for wrist pins, kingpins requiring high hardness and fatigue strength.

6.3 Nickel-Chromium Steels (3XXX)

Nickel chromium steels were developed as an alternate to nickel steels. Partial substitution of chromium for nickel gives properties of the nickel steels with lower cost. Nickel gives toughness and chromium hardens the alloy. Care has to be taken while adding chromium and nickel since they have opposite effect on the eutectoid (austenitizing) temperature. Normally the ratio between nickel and chromium is maintained below 2.5. Above this ratio, narrows the austenitic range and hence heat treatment will become critical.

6.4 Molybdenum Steels (4XXX)

In general, molybdenum is used to enhance the properties imparted by other alloying elements especially nickel and chromium as a synergic effect. These steels normally have molybdenum in the range of 0.15–0.6%. Addition of molybdenum resists softening during tempering and hence can be used for slightly elevated temperature applications. It increases ductility, toughness, machinability and hardenability. The major advantage is that it eliminates temper brittleness. Molybdenum containing steels require high temperature and longer heating time for austenitizing since the dissolution of MoC in austenite is slow. Among the various grades of molybdenum steels, 41XX is more popular and used in pressure vessels. The grades 43XX,

46XX, 47XX, and 48XX have nickel in addition to molybdenum. Due to nickel, these grades have good ductility and toughness.

6.5 Chromium Steels (5XXX)

Addition of chromium increases hardenability, strength, hardness, and wear resistance. However, the ductility is reduced. Chromium steels, if tempered above 540°C suffer temper embrittleness. The case carburized chromium steels have hard case but tough core is missing unlike nickel steels.

6.6 Vanadium Steels (6XXX)

Like molybdenum, vanadium is used to enhance the properties imparted by other alloying elements especially chromium. But the vanadium carbides are stronger than molybdenum carbides. These grades of steels normally have vanadium in the range of 0.15–0.6%. Vanadium carbides are insoluble in normal austenitizing temperature and retards the grain growth of austenite. Like molybdenum steels, vanadium steels also resist softening and have good creep resistance.

6.7 Tungsten Steels (7XXX)

Tungsten was the first among the alloying elements systematically used as early as in the middle of the 19th century to improve steel properties. Tungsten forms very stable carbides. These grades of steel have both wear resistance and temperature resistance. Importance of tungsten in steel has steadily increased, with the steel industry being the largest tungsten consumer. The use of tungsten in structural steels declined since 1940 because alloying with molybdenum and chromium as well as with vanadium and nickel has given better performance at lower cost. Today, the series itself is removed from the list like nickel steels. Although tungsten is not popular alloying element in low alloy steels, tungsten is the main alloying element in tool and die steels due to its excellent properties.

6.8 Triple Alloy Steels (8XXX and 9XXX)

Triple alloys are hardenable chromium, molybdenum, and nickel low alloy steels often used for carburizing to develop a case hardened part at cheaper cost. It is a balanced alloy and has good welding qualities. Case hardening of this alloy will result in good wear characteristics. The nickel imparts good toughness and ductility while the chromium and molybdenum contribute increased hardness penetration and wear. These grades are used where a hard, wear-resistant surface and a ductile core is required. In a few grades, lead is added to have better machinability and surface finish. These grades are available in aircraft quality and bearing quality.

6.9 Silicon Steels

Silicon steels are also called as electric steels. Silicon promotes a ferritic microstructure and increases strength. Silicon reduces the hysteresis loss in electrical applications. About 3% silicon is used in these grades of steels and worldwide researchers are working to make electric steels with about 6% silicon, which will have high magnetic properties. Silicon also increases resistance to oxidation and used for moderate temperature applications. Care must be taken while using these grades of steel at high temperatures because silicon is a graphatizer and causes graphatization of cementite.

6.10 Boron Steels

Like other alloying elements, boron has no effect on Ms temperature and retained austenite. It does not change the fineness of pearlite, nor it produces any solid solution strengthening in ferrite. It shows no effect on tempering response but the outstanding feature of boron is the improvement in hardenability by the addition of even a minute (0.0015% to 0.0030%) quantity of boron. In case an excessive amount of boron (greater than 0.0030%) is present, the boron constituents become segregated in the austenite grain boundaries, which not only lowers hardenability but also may decrease toughness, cause embrittlement and produce hot shortness.

Boron delays the beginning of the transformation of ferrite by suppressing the nucleation of proeutectoid ferrite on austenitic grain boundaries but it has no effect on growth, that is, end of transformation, which facilitates isothermal processes such as ausforming to form tough bainite. In addition, it increases hardenability by preventing the formation of softer structures during cooling from the austenization temperature, after annealing or hot working. The effect of boron on hardenability also depends on the amount of carbon in the steels. The effect of boron is inversely proportional to carbon. Boron is much more effective at low carbon levels. Boron contribution falls to zero as the eutectoid carbon content is approached. An interesting sidelight of this phenomenon is that the carburized boron steels are marked by a high core hardenability but a low case hardenability. Boron has long been used as a replacement for other alloying elements in heat treatable steels, especially when these constituents were in short supply. Boron does not retard grain growth but reduces grain boundary diffusion hence the creep properties. Boron steels are becoming increasingly popular and their application is becoming more diverse. Their higher properties, at a reasonable price, are achieved through advanced manufacturing technology. Although boron steels were originally designed mainly for the hard, wear resistant elements, now they are also being used for other applications.

Boron is not added as a separate alloying element in plain carbon steels. The letter B is introduced after the first two digits. For example, 51BXX indicates that the steel has about 0.8% chromium along with boron.

Boron steels are used when the base composition meets mechanical property requirements (toughness, wear resistance, etc.), but hardenability is insufficient for the intended section size. Carbon-manganese-boron (C-Mn-B) steels are generally specified as replacements for alloy steels for reasons of cost as C-Mn-B steels are far less expensive than alloy steels of equivalent hardenability. Boron is sometimes used in non-heat-treated steels as of nitrogen scavenger. By avoiding interstitial nitrogen, boron makes the steel more formable and eliminates the need for strain age suppressing anneals especially in steels used for automotive strip stock.

Boron is added to some steels for the nuclear industry since it has a high neutron absorption capability. Levels of 4% boron or more have been used, but due to the lack of hot ductility and weldability in these steels, boron contents of 0.5% to 1.0% are more common for neutron absorption applications. For this application of boron, the Fe-B has to be of the highest purity. Great care has to be taken while melting the boron steels since boron reacts readily with oxygen and nitrogen and is not useful to steels if it is in combined form. Boron must be in its atomic state to improve hardenability. If this precaution is not taken then it can lead to erratic heat treatment response. Care has to be taken while heat treating since boron may also become ineffective if its state is changed by incorrect heat treatment. For example, high austenitizing temperatures must be avoided as well as temperature ranges where certain precipitation of boron occurs. Also, while heat treating, oxidizing and nitriding atmospheres are to be avoided. Boron steels are not to be carbonitrided.

Typical mechanical properties of most commonly used low alloy steels are listed in Table 6.1

6.11 Interstitial Free Steels

The low carbon steels (0.2% C) combine moderate strength with excellent ductility and are used extensively for their fabrication properties in the

TABLE 6.1

Mechanical properties of selected low alloy steels

S. No	AISI	Condition	UTS (MPa)	YS (MPa)	Elongation %	Hardness (HB)
1	**1010**	normalized	324	179	28	95
		cold drawn	365	305	20	105
2	**1020**	as rolled	448	346	36	143
		normalized	441	330	35	131
		annealed	393	294	36	111
3	**1030**	normalized	490	260	21	75 HRB
		hardened and tempered	540	440	12	149
4	**1040**	hot rolled	496	290	18	144
		cold drawn	552	490	12	160
		hardened and tempered at 705°C	607	421	33	183
		hardened and tempered at 204°C	779	600	19	262
5	**1050**	hot rolled	620	338	15	180
		cold drawn	690	579	10	200
		hardened and tempered at 705°C	662	421	30	192
		hardened and tempered at 204°C	986	758	10	321
6	**1080**	as rolled	1010	586	12	293
		normalized	965	524	11	293
		annealed	615	375	24	174
7	**1117**	hot rolled	430	230	23	121
		cold rolled	480	400	15	137
8	**1213**	hot rolled	379	228	25	110
		cold rolled	517	340	10	150
9	**1340**	annealed	703	434	26	207
		hardened and tempered at 705°C	690	517	25	235
		hardened and tempered at 538°C	993	910	17	363
		hardened and tempered at 370°C	1520	1360	13	461
		hardened and tempered at 204°C	1960	1610	8	578
10	**3140**	annealed	655	462	25	187
		hardened and tempered at 705°C	792	648	23	233
		hardened and tempered at 538°C	1050	920	17	311

(*Continued*)

TABLE 6.1 (Cont.)

S. No	AISI	Condition	UTS (MPa)	YS (MPa)	Elongation %	Hardness (HB)
		hardened and tempered at 370°C	1520	1380	13	461
		hardened and tempered at 204°C	1930	1710	11	555
11	**4130**	annealed	558	359	28	156
		hardened and tempered at 705°C	676	614	28	202
		hardened and tempered at 538°C	986	910	16	302
		hardened and tempered at 370°C	1430	1240	13	415
		hardened and tempered at 204°C	1610	1360	12	461
12	**4140**	annealed	655	414	26	197
		hardened and tempered at 705°C	807	690	23	235
		hardened and tempered at 538°C	1160	1050	17	341
		hardened and tempered at 370°C	1590	1460	13	461
		hardened and tempered at 204°C	2000	1730	11	578
13	**4150**	annealed	731	379	20	197
		hardened and tempered at 705°C	880	800	20	262
		hardened and tempered at 538°C	1360	1250	11	401
		hardened and tempered at 370°C	1700	1580	10	495
		hardened and tempered at 204°C	2070	1710	10	578
14	**4340**	annealed	745	469	22	217
		hardened and tempered at 705°C	965	827	23	280
		hardened and tempered at 538°C	1180	1090	16	363
		hardened and tempered at 370°C	1590	1420	12	461
		hardened and tempered at 204°C	1950	1570	11	555
15	**5140**	annealed	572	290	29	167
		hardened and tempered at 705°C	717	572	27	207

(Continued)

TABLE 6.1 (Cont.)

S. No	AISI	Condition	UTS (MPa)	YS (MPa)	Elongation %	Hardness (HB)
		hardened and tempered at 538°C	1000	896	18	302
		hardened and tempered at 370°C	1520	1380	11	429
		hardened and tempered at 204°C	1900	1560	7	534
16	**5150**	annealed	676	359	22	197
		hardened and tempered at 705°C	800	700	22	241
		hardened and tempered at 538°C	1100	1030	15	321
		hardened and tempered at 370°C	1650	1520	10	461
		hardened and tempered at 204°C	2150	1720	8	601
17	**5160**	annealed	724	276	17	197
		hardened and tempered at 705°C	793	690	23	229
		hardened and tempered at 538°C	1170	1040	14	341
		hardened and tempered at 370°C	1810	1630	9	514
		hardened and tempered at 204°C	2220	1790	4	627
18	**6150**	annealed	662	407	23	197
		hardened and tempered at 705°C	814	738	21	241
		hardened and tempered at 538°C	1260	1190	12	375
		hardened and tempered at 370°C	1700	1540	10	495
		hardened and tempered at 204°C	2170	1860	7	601
19	**8650**	annealed	717	386	22	212
		hardened and tempered at 705°C	841	779	21	255
		hardened and tempered at 538°C	1210	1070	14	363
		hardened and tempered at 370°C	1650	1530	12	495
		hardened and tempered at 204°C	1940	1720	11	555

(Continued)

TABLE 6.1 (Cont.)

S. No	AISI	Condition	UTS (MPa)	YS (MPa)	Elongation %	Hardness (HB)
20	**8740**	annealed	690	414	22	201
		hardened and tempered at 705°C	820	690	25	241
		hardened and tempered at 538°C	1210	1150	15	363
		hardened and tempered at 370°C	1570	1460	12	461
		hardened and tempered at 204°C	2000	1650	10	578
21	**9255**	annealed	780	490	22	229
		hardened and tempered at 705°C	896	703	21	262
		hardened and tempered at 538°C	1250	1100	14	352
		hardened and tempered at 370°C	1790	1650	5	534
		hardened and tempered at 204°C	2140	1980	2	601

annealed or normalized condition for bridges, buildings, cars and ships. However, even 0.2% carbon, has limited ductility for deep drawing operations, and brittle fracture becomes a problem, particularly for welded thick sections. Improved low carbon steels < 0.2% are produced by deoxidising or "killing" the steel with aluminium or silicon or by adding manganese to refine the grain size. It is now more common, however, to add small amount of (< 0.1%) of niobium, which reduces the carbon content by forming NbC particles. These particles not only restrict grain growth but also give rise to strengthening by precipitation hardening within the ferrite grains. Other carbide formers, such as titanium, may be used but because niobium does not deoxidize, it is possible to produce a semikilled steel ingot having reduced ingot pipe, giving increased tonnage yield per ingot cast. These types of steels are called Interstitial Free (IF) steels. Due to their excellent formability and nonaging characteristics are widely preferred material for automotive applications.

From early 1970s, carbon contents close to 0.01% became readily available and it was then necessary to add sufficient alloying addition (titanium and niobium) to be able to combine with all the carbon and nitrogen in the steel and to leave a small surplus in order to obtain very good formability. More recently, with improvements in vacuum degassing techniques, ultra-low carbon (ULC) contents below 0.003% have become easily available and with these ultra-low carbon contents, it has been possible to lower the alloy addition (close to 0.01% of niobium) while still achieving high formability. However, a number of different combinations of niobium and titanium are also in use.

6.12 Applications as Engineering Material

As nickel has the ability to impart high toughness, especially at low temperatures, nickel steels (2XXX) led to the development of cryogenic steels having important applications in the transportation and storage of liquefied gases. A structural steel for lower service temperature without the risk of brittle failure must contain more nickel. Thus, a low carbon 2.5% nickel steel can be used down to –60°C while 3.5% nickel lowers the allowable temperature to –100°C and 9% Ni steels are useable up to –196°C. The nickel-chromium steels (31XX) have good properties at low cost and used for drive axle shafts, connecting rods. The 33XX are used for heavy duty applications since they have good hardenability. Among the various grades of molybdenum steels, 41XX is more popular and used in pressure vessels. The grades 43XX, 46XX, 47XX and 48XX are used mainly where high fatigue and tensile properties are required. Chromium steels with high carbon is used for making knifes and for keen cutting edges. The 52100 grade is used for making rollers of antifriction bearings due to high hardness and wear resistance. Vanadium steels are used when the molybdenum steels could not with stand the temperature or stress. Triple alloy steels are used as gears, ring gears, shafts, pinions, spline shaft, piston pins, oil pumps, piston rods and liners, cams, oil tool slips, jigs, gauges, plastic molds, jaws, and crankshafts or similar applications. The main application of silicon steels is in making sheets for transformer cores and for magnetic applications since the hysteresis loss is minimum in silicon steels. Silicon steels are also used in automobile valves called as valve steels having 3–4% silicon since it is resistant to oxidation, both at high temperatures and in strongly oxidizing solutions at lower temperatures. Boron steels are used primarily in punching tools, spades, knives, saw blades, and safety beams in vehicles, and so on, where high hardenability is required. Boron is also used in deep drawing steels, where it removes interstitial nitrogen and allows lower hot rolling temperatures. Applications for carbon-manganese-boron steels include earth scraper segments, track links, rollers, drive sprockets, axle components and crankshafts. Interstitial free steels are used in making automotive bodies due to its excellent deep drawing properties.

The other categories of low alloy steels are high strength low alloy steels (HSLA), dual phase steels, and transformation induced plasticity (TRIP) steels. They also have low alloying elements, but the formulation and treatment are different. The metallurgy of these alloy steels are dealt in subsequent chapters. Stainless steels, maraging steels, and tool steels fall under the category of high alloys steels because they have alloying elements more than 10%. The physical metallurgy aspects of these steels are also described in later chapters of this book.

7

High Strength Steels

7.0 Introduction

The low strength steels considered in the previous chapter have been in use for many years since they exhibit a number of useful properties such as very good formability, sufficient strength, good weldability, corrosion resistance or suitable for corrosion protection, good surface quality, and so on. In the last decade, a clear interest for the development of high strength steels has emerged. With stronger steels, thinner parts can be fabricated and weight savings can be obtained. Within the family of low carbon steels, several variants like Al killed steel, interstitial free (IF) steel, and so on, have been developed and were discussed in the previous chapter.

7.1 Definition and Need of High Strength Steels

The definition was framed by automotive manufacturers as they are the main users. The definition for high strength steels (HSS) and advanced high strength steels (AHSS) is arbitrary because the terminology used to classify steel products varies considerably throughout the world. It is generally accepted that the transition from mild steel to high strength steels (HSS) occurs at a yield strength of about 270 MPa. For yield strength levels between 280 to 350 MPa, typically a simple carbon manganese (C-Mn) steel is used. The composition of these steels is similar to low carbon mild steels, except they have more carbon and manganese to increase the strength to the desired level. This is also known as conventional high strength steels. This approach usually is not practical for yield strengths greater than 350 MPa due to drop off in elongation and weldability. The yield strengths between 280 and 550 MPa are achieved in high strength low alloy (HSLA) steels, also known as microalloyed (MA) steels. This family of steels usually has a microstructure of fine grained ferrite that has been strengthened with carbon and/or nitrogen precipitates of titanium, vanadium, or niobium (columbium). Adding manganese, phosphorus, or silicon further increases the strength. These steels can be formed successfully when users are aware of the limitations of the higher strength, lower formability trade off.

Steels with yield strength levels in excess of 700 MPa are called as advanced high strength steels (AHSS) and steels with tensile strengths exceeding 780 MPa are called "ultrahigh-strength steels." AHSS with tensile strength of at least 1000 MPa are often called "Giga Pascal steel" (1000 MPa = 1GPa) as represented in Figure 7.1. AHSS have excellent strength combined with excellent ductility, and thus meet many functional requirements. Dual phase (DP), transformation induced plasticity (TRIP), and martensitic steels are some of the grades collectively referred to as AHSS.

The following two grades of steels are not classified under HSS or AHSS: (i) Austenitic stainless steel having strength level of above 700 MPa with elongation of about 60% and (ii) maraging steels with strength above 1900MPa and about 40% elongation,. These steels due to high alloying content are expensive choices for many components especially in automobile sector and are not normally preferred for automobile applications. The austenitic grade stainless steel is used where good corrosion resistance is required and maraging steels are candidate material for aerospace applications and are discussed in Chapter 8.

The need for HSS and AHSS, mainly in the automobile market, arose due to the oil crises of 1973 and 1979. The oil crisis provided the initial stimuli for weight reduction. The car body is assembled from large body panels and constitutes about 25–30% of the total weight of a medium size car. It is, therefore, the heaviest vehicle component. Therefore reducing the

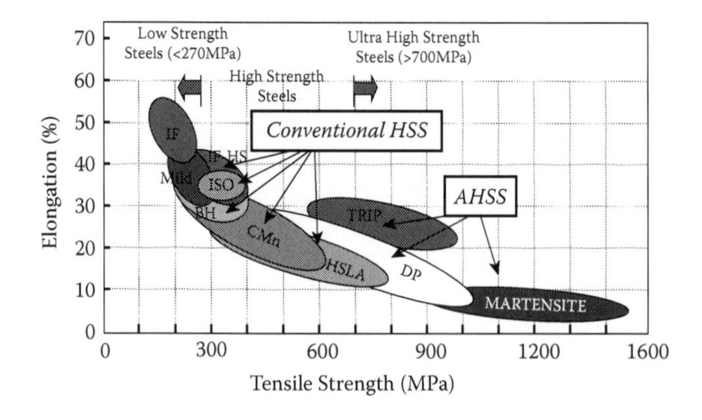

FIGURE 7.1
Representation of low, high, and ultra-high strength steels

weight of the body could have a significant impact on the total weight of the complete car. It was planned that this could be done by the substitution of steel by aluminium or plastics, but the penalty would have been increased cost. Subsequently, it was generally accepted that the use of high strength steel, which would enable component performance to be maintained at a reduced thickness, was the only way to achieve both weight and cost savings. Although the initial application is in automobile industries, the HSS has captured the market in other engineering fields also.

7.2 Types and Problems in Developing High Strength Steels

The first high strength steels were the mild steels with a higher degree of temper rolling to increase the yield strength. The main problem, however, was the low formability which restricted their use. Nevertheless, strengthening by cold reduction is an effective way of strengthening when a very low degree of formability is acceptable. Cold reduced steels continue to be used, therefore, for many strapping applications and in the zinc coated condition for corrugated roofing panels. Attempts made to increase the strength and ductility (formability) in steels led to the development of following grades of steels.

7.2.1 Modified Standard Low Alloy Steels

The modification was carried out to AISI 4340 and equivalent medium carbon nickel, chromium, and molybdenum steels. These steels can be oil-quenched and tempered to any desired strength and hardness. Properties can be improved by vacuum arc or electroslag remelting to lower the hydrogen, nitrogen, and oxygen contents and to reduce the number of nonmetallic inclusions. An improvement is obtained by increasing the silicon content of 4340 from 0.15–0.30% to 1.45–1.80% to take advantage of the well-known effect of silicon in inhibiting the growth of carbide particles. This steel, called 300 M, can be tempered at a higher temperature than 4340 to develop the same hardness, thereby reducing quenching stresses. The addition of silicon also moves the temper embrittlement range to higher temperatures. Tempering at 315°C produces a maximum in notch impact toughness and near maximum yield strength. The additional silicon also increases hardenability and adds a component of solid solution strengthening. A small addition of vanadium may provide fine V(C, N) particles during tempering. Hy-Tuf is a modification of AISI 4130 with increased manganese and silicon with an addition of nickel. In D6AC the chromium and molybdenum contents are increased over those in 4340. These steels are also called as martensitic steels (MS or Mart). Serious limitations in producing these conventional high strength steels are: (i) the associated reduction in fracture toughness and hydrogen embrittlement. Carbon is one of the strongest elements that can increase the strength by forming martensite but it affects the toughness. Martensite is more susceptible to hydrogen embrittlement and also has problems related with fabrication – forming, welding, and so on; (ii) the alloying elements increase strength but increase austenitizing temperatures and allied problems during heat treatment; (iii) the conventional high strength steels have limited formability because their high strength is developed prior to the forming process; and (iv) attention must be paid when the high strength steels are electroplated, as there have been many failures of high strength steels into which hydrogen was introduced during electroplating of protective surface layers. Concentration of a few parts per million is often sufficient to cause failure. Although much hydrogen escapes from steel in the molecular form during treatment, some can remain and precipitate at internal surfaces such as inclusion/matrix and carbide/matrix interfaces, where it forms voids or cracks.

Hence, it is necessary in high strength steels to reduce carbon as low a level as possible, consistent with good strength. Developments in the technology of high alloy steels have produced high strengths in steels with very low carbon contents. The loss in strength due to carbon is compensated by other strengthening mechanisms such as substitutional solid solution, grain refinement, precipitation hardening, and so on. The susceptibility to hydrogen is overcome by avoiding vulnerable martensitic structure, as high strength steels have more resistant bainitic, ferritic, and spherodized microstructure by thermomechanical treatments and controlled heat treatments.

7.2.2 Rephosphorized Steel

The main emphasis was the development of steels in which the loss in ductility with increasing strength was minimized. Substitutional solid solution strengthened steels were developed in which the loss in elongation per unit increase in strength is less for a solid solution strengthened steel. The first possibility to increase the strength of an alloy is to increase the content of atoms of solid solution elements such as phosphorus, silicon and manganese which will strengthen the steel. But the last two will reduce the formability to an unacceptable level.

However, the strength increase that could be obtained with a rephosphorized addition was limited by the detrimental effect of phosphorus on welding. Hence only small amounts of phosphorus (between 0.04 and 0.08 %) can be tolerated which will increase the strength roughly by 40–80 MPa. These steels were restricted to relatively modest strength increases with minimum yield stresses up to 300 MPa. The steels were highly formable. Limited amounts of phosphorus were added to IF steel. These grades called IF-HSS (high strength steels) have an ultimate tensile strength (UTS) of 360–400 MPa and good formability.

7.2.3 Micro Alloyed or High Strength Low Alloy (HSLA) Steels

By the middle of 1960s, the higher strength steels, based on micro alloy additions, had become established and each of the classification societies introduced specifications with yield stress values as a substitute to cold rolled high strength steels. The higher strengths are achieved by grain refinement and precipitation strengthening while the carbon content is maintained as low as possible and also the carbon equivalent (CE) is restricted to 0.41 % max for ease of welding and forming. Grain refinement and precipitation hardening is achieved through the addition of small amounts of carbide or carbonitride forming elements like niobium and titanium. During hot rolling, undissolved fine particles restrict the grain growth of austenite grains. In the last hot rolling passes, austenite will not recrystallize anymore and the grains will be flattened out. During subsequent cooling many ferrite nuclei will be activated and will provide a fine ferrite grain size, which is a first important contribution to strengthening. Additional precipitates are also formed which can further strengthen the ferrite. A typical HSLA steel can have a yield strength (YS) of about 320 MPa and an ultimate tensile strength (UTS) of 440 MPa, but the formability is too low for car body panels. HSLA steels can however be used for structural applications, for example, beams. Various types of HSLA were developed based on the needs and requirement.

7.2.4 Bake Hardening (BH) Steels

The steel must be relatively soft and highly deformable to make in to desired shape. However, after forming, the part should be as strong as possible. The yield point phenomenon and strain aging help to solve this dilemma by strengthening the material during strain aging. When a plain low carbon steel is strained plastically to a particular strain showing serrations (region A of Figure 7.2) called as yield point phenomenon due to locking of moving dislocations by the interstitial carbon or nitrogen atoms. If it is unloaded and reloaded again without any appreciable time delay or any heat treatment yield point does not occur since the dislocations have been torn away from the carbon and nitrogen atoms (region B of Figure 7.2). If it is unloaded and reloaded after aging for several days at room temperature or several hours at slightly elevated temperatures (120–170°C), then yield point reappears and moreover, the yield point will be increased (region C of Figure 7.2). This is due to diffusion of carbon and nitrogen atoms to dislocations during the aging to form new solute atmospheres anchoring the dislocations.

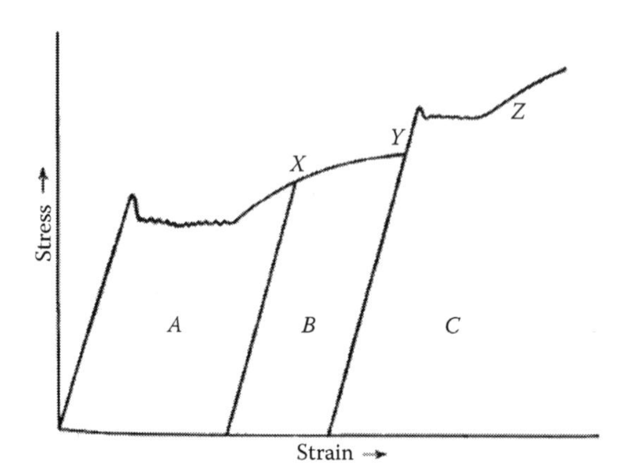

FIGURE 7.2
Loading and unloading of plain low carbon steels showing yield point phenomenon and strain aging.

This phenomenon is used in bake hardening (BH) steels. The steel is formed in the region B and so-called "paint baking cycle," is carried out at about 170°C for 20–30 minutes especially in car body manufacturing leading to higher yield strength (point Z in Figure 7.2). This increases the yield strength roughly by 50 MPa after paint baking cycle. The important caution is to eliminate strain aging during forming (region A in Figure 7.2) because the yield point can lead to difficulties with surface markings or stretcher strains due to localized heterogeneous deformation. The steels are subjected to about 2% reduction after the final pass of the rolling (skin pass rolling), which eliminates yield point serration and push the steel to region B so that the yield point phenomenon (region A in Figure 7.2) will not occur while making the actual components. Nitrogen plays a more important role in the strain aging of iron than carbon because it has a higher solubility and diffusion coefficient and produces less complete precipitation during slow cooling.

7.2.5 Dual Phase (DP) and Transformation Induced Plastic (TRIP) Steels

Dual phase steels are low carbon steel grades with carbon content around 0.1%. These steels have soft ferrite and 10–20% dispersed hard martensite islands giving tensile strengths up to and above 600 MPa. This microstructure is obtained by soaking in the intercritical range (~800°C), followed by rapid cooling. During intercritical annealing, part of the austenite transforms into ferrite. DP steels have relatively high strain hardening values for their strength but the formability is low. One way to increase formability is to have small amount of austenite, which on straining is converted as martensite contributing to higher strength and formability. Based on this line of thought, transformation induced plasticity (TRIP) steels were developed, which have some retained austenite in addition to the dual phase (ferrite + martensite).

7.2.6 High Strength Steels by Thermo Mechanical Treatments

There is a limit to the strengthening that can be achieved by a combination of grain refinement, solid solution and precipitation effects. Thermomechanical processing has permitted the development of high strength steels with low carbon contents and this has contributed greatly to improved weldability and, therefore, employed to obtain values of tensile strength above 500 MPa, depending on whether the steel is in the hot rolled condition or has also been cold rolled and annealed.

7.2.6.1 Patenting of Steel Wires

Steel wires are used extensively throughout the world for many critical engineering applications such as high strength cables for bridges, cable and ski lifts, general haulage, for example, ship moorings, and, on a large scale, for reinforcing radial tires. They are also widely used as piano and violin strings. In all cases, their properties of very high strength and toughness are just about unique.

The high strength steel wire is basically a fully eutectoid 0.8%C steel containing some manganese and silicon designed to develop the maximum amount of pearlite, that is, about 100%. The wire is strengthened by a process called patenting. To produce fine wire it starts as coiled rod of diameter around 5 mm. In this condition, the strength is about 1100–1200 MPa and the aim of the patenting process is to multiply this by a factor of about 3. The rod is first given a preliminary drawing reduction down to 1–2 mm diameter without lubrication. It is then passed into a furnace and heated in the range 950–1000°C, that is, in the single phase austenite domain for transformation into homogeneous austenite with all the carbon in solid solution. It is then very carefully cooled to transform into fine pearlite. This can be done isothermally by cooling in liquid lead or in a fluidised bath or even by controlled air cooling and then drawn into thin wire. The resulting work hardened pearlite is extremely strong – probably the strongest material known possessing some ductility and therefore toughness.

The cooling rate is one of the critical parameters in the process. Also during cooling the austenite to pearlite, the exothermic transformation gives off sufficient heat to increase the temperature in the range of transformation. The cooling rate allows for the heat of transformation to follow a finely judged path into the pearlite domain that leads to the formation of a very fine pearlite, that is, in the range of 500–600°C (and without any bainite). The microstructure is then basically 100% pearlite with a spacing of about 0.25 micron.

Patented steel wires are usually drawn up to strengths of about 3000 MPa for use in tyre reinforcements. However, there is a trend to develop even higher strength wires by increasing the drawing strain and, in some cases, the carbon content. These wires are aimed to attain 4000 MPa.

7.2.6.2 Severe Plastic Deformation of Steels

Patenting is limited to only thin wires and cables. Thick sections or sheets could not be made by patenting. Alternatively, severe plastic deformation methods, especially equal channel angular pressing (ECAP) is under research to develop high strength coupled with ductility in low carbon steels used in the construction industry. Table 7.1 shows the properties achieved by the low carbon steel (Fe 450 grade) after ECAP.

In ECAP, true shear strains more than 1 is possible. At these levels of strain, and even before, the cementite layers break up and, surprisingly begin to dissolve locally so that excess carbon goes into solution in the ferrite. And it is possible to obtain duplex microstructure of ferrite (about 13%) and martensite (about 87%) in the ECAPed low carbon steels upon quenching from intercritical temperature as shown in Figure 7.3 achieving tensile strength of 638 MPa and 28% elongation. These types of steels show better combination of strength and ductility than the as ECAPed samples due to dual phase structure of ferrite and martensite. In the dual phase microstructure, the ferrite contributes to the maximum elongation and the martensite contributes to the strength. However, this behavior is in the research level only and becoming popular now.

These thermomechanical processes have limitation in industrial scales for the following reasons:

TABLE 7.1

Tensile strength and % elongation of annealed, as received and as ECAPed mild steel (Fe 450 grade)

S.No	Sample condition	Tensile strength (MPa)	% elongation
1	**as received**	459	26
2	**Annealed**	325	51
3	**ECAP at 200°C**	683	13
4	**ECAP at 250°C**	649	12
5	**ECAP at 300°C**	630	13
6	**ECAP at 350°C**	619	13
7	**ECAP at 200°C – 2 passes**	707	12
8	**Intercritical annealed**	638	28

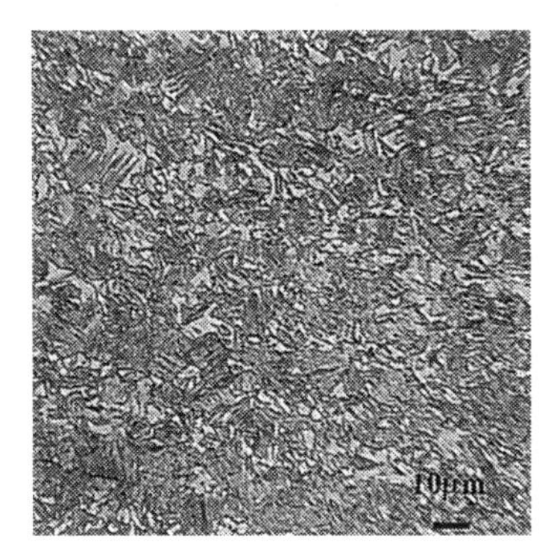

FIGURE 7.3
Microstructure of quenched ECAPed steel from intercritical temperature showing ferrite and martensitic structure.

1. Large reductions in cross section must be made, so large sections cannot be treated.
2. At the temperatures involved, loads on equipment will be severe.
3. The steel must be used in the shape in which it is formed since the fabricability is limited.
4. Joining is difficult; fusion welding is almost impossible.

7.2.7 Nanostructured Steels

It has been possible for some time to produce iron in which the space filling crystals are just 20 atoms wide. These samples were prepared from the vapour phase followed by consolidation. Although of limited engineering value, work of this sort inspired efforts to invent methods of making large quantities of steels with similar fine grain structures. Modern technologies allow steels to be made routinely and in large quantities with grain sizes limited to a minimum of about 1 micron by recoalescence effects. Processes generally involving severe thermomechanical processing have been developed to achieve nanostructured ferrite grains in steel, in the range of 20–100 nm. Although the nanostructured steels are strengthened as expected from the Hall–Petch equation, they tend to exhibit unstable plasticity after yielding. The plastic instability occurs in both tension and in compression testing, with shear bands causing failure in the latter case. Also, the steels lack ductility due to nanograins, which diminish work hardening capacity. The conventional mechanisms of dislocation multiplication fail in nanograined steels because of the proximity of the closely spaced boundaries which then become impossible to accumulate dislocations during deformation. Grain boundaries are also good sinks for defects. These effects cause nanostructured materials not to work harden.

7.2.8 Bainitic Steels

There are large markets for steels with strengths less than 1000 MPa, and where the total alloy concentration rarely exceeds 2 %. Bainitic steels are well suited for applications within these constraints. The bainitic structure has better hardness, strength and ductility comparable to hardened and tempered martensite. If a steel is cooled rapidly from an austenite state to a temperature close to 450°C, the ferrite and pearlite reactions are suppressed and the austenite will transform to a lower temperature transformation product called bainite and is called austempering explained in the previous chapter. However, alloy design must be careful in order to obtain the right microstructures. Steels with inadequate hardenability tend to transform to mixtures of allotriomorphic ferrite and bainite. Attempts to improve hardenability usually lead to partially martensitic microstructures. The solution therefore lies in low alloy, low carbon steels, containing small amounts of boron and molybdenum to suppress allotriomorphic ferrite formation. Boron increases the bainitic hardenability. Other solute additions can, in the presence of boron, be kept at sufficiently low concentrations to avoid the formation of martensite. A typical composition (in wt%) might be Fe–0.1C–0.25Si–0.50Mn–0.55Mo – 0.003B. Steels like these are found to transform into virtually fully bainitic microstructures with very little martensite using normalizing heat treatments. The reduced alloy concentration not only gives better weldability, but also a higher strength due to the refined bainitic microstructure. These steels are made with very low impurity and inclusion contents, to avoid the formation of cementite particles. Some typical alloy compositions are given in Table 7.2.

Medium strength steels with the same microstructure but somewhat reduced alloy content have found applications in the automobile industry as crash reinforcement bars to protect against sidewise impact. Another major advance in the automobile industry has been in the application of bainitic forging alloys to manufacture components such as cam shafts. These were previously made of martensitic steels by forging, hardening, tempering, straightening and finally stress relieving. All of these operations are now replaced by controlled cooling from the die forging temperature, to generate the bainitic microstructure, with cost savings which on occasions have made the difference between profit and loss for the entire unit.

By inoculating molten steel with controlled additions of nonmetallic particles, bainite can be induced to nucleate intragranularly on the inclusions, rather than from the austenite grain surfaces. This intragranularly nucleated bainite is called "acicular ferrite." It is a much more

TABLE 7.2

Chemical composition, wt%, of typical bainitic steels

Alloy	C	Si	Mn	Ni	Mo	Cr	V	B	Nb	Other
Early bainitic steel	0.10	0.25	0.5	–	–	0.003	–	–	–	–
Ultra-low carbon	0.02	0.20	2.0	0.3	0.30	–	–	0.01	0.05	–
Ultra-high strength	0.20	2.00	3.0	–	–	–	–	–	–	–
Creep resistant	0.15	0.25	0.5	–	1.00	2.30	–	–	–	–
Forging alloys	0.10	0.25	1.0	0.5	1.00	–	–	–	0.10	–
Inoculated	0.08	0.20	1.4	–	–	–	–	–	0.10	0.012 Ti
Nanostructured	1.00	1.50	1.9	–	0.26	1.26	0.1	–	–	–

disorganized microstructure with a larger ability to deflect cracks. Inoculated steels are now available commercially and are being used in demanding structural applications such as the fabrication of oil rigs for hostile environments.

Advances in rolling technology have led to the ability to cool the steel plate rapidly during the rolling process, without causing undue distortion. This has led to the development of "accelerated cooled steels," which have a bainitic microstructure that are highly formable and compete with conventional control rolled steels.

It would be nice to have a strong material with good toughness without requiring mechanical processing or rapid cooling to reach the desired properties. The following conditions are required to achieve this:

(i) The material must not rely on perfection to achieve its properties. Strength can be generated by incorporating a large number of defects such as grain boundaries and dislocations, but the defects must not be introduced by deformation if the shape of the material is not to be limited.

(ii) Defects can be introduced by phase transformation, but to disperse them on a sufficiently fine scale requires the phase change to occur at large under cooling (large free energy changes). Transformation at low temperatures also has the advantage that the microstructure becomes refined.

(iii) A strong material must be able to fail in a safe manner. It should be tough.

(iv) Recalescence (a temporary rise in temperature during cooling of a metal, caused by a change in crystal structure) limits the undercooling that can be achieved. Therefore, the product phase must be such that it has a small latent heat of formation and grows at a rate which allows the ready dissipation of heat.

Recent discoveries have shown that carbide free bainite can satisfy these criteria. Bainite and martensite are generated from austenite without diffusion by a displacive mechanism. Not only does this lead to solute trapping but also a huge strain energy term, both of which reduce the heat of transformation. The growth of individual plates in both of these transformations is fast, but unlike martensite, the overall rate of reaction is much smaller for bainite. This is because the transformation propagates by a sub unit mechanism in which the rate is controlled by nucleation rather than growth. This diminishes recalescence. This leads to the development of nanostructured bainitic steels.

7.2.8.1 Nanostructured Bainitic Steels

The major setback in the nanograined material is the lack of ductility due to the loss of work hardening capacity. This can be resolved by introducing retained austenite in the microstructure. The strain or stress induced martensitic transformation of this austenite enhances the work hardening coefficient, making it possible to get substantial ductility in nanostructured steels. However, the amount of austenite must be above a threshold level, which is estimated to be about 10 vol. %. The same methodology is adopted in nanostructured bainitic steels. However, there is a possibility of formation of coarse cementite particles, which are detrimental for toughness. The precipitation of cementite during bainitic transformation can be suppressed. This is done by alloying the steel with about 1.5 % of silicon, which has a very low solubility in cementite and greatly retards its growth. An interesting microstructure results when this silicon alloyed steel is transformed into upper bainite. The carbon that is rejected into the residual austenite, instead of precipitating as cementite, remains in the austenite and stabilizes it down to ambient temperature. The resulting microstructure consists of fine plates of bainitic ferrite separated by carbon enriched regions of austenite.

With these precautions an alloy has been designed with the approximate composition Fe–1C–1.5Si–1.9Mn–0.25Mo–1.3Cr–0.1V %, which on transformation at 200°C, leads to bainite plates that are only 20–40 nm thick separated by carbon enriched films of retained austenite. This is the hardest ever bainite, that can be manufactured in bulk form, without the need for rapid heat treatment or mechanical processing. It is important to note that it consists only of two phases, slender plates of bainitic ferrite in a matrix of carbon-enriched austenite.

The slender plates of bainite are dispersed in stable carbon enriched austenite, with its face centred cubic lattice, buffers the propagation of cracks. The bainite obtained by transformation at very low temperatures is the hardest ever (700HV, 2500 MPa), has considerable ductility (5–30%) and is tough (30–45 MPa $m^{1/2}$) and does not require mechanical processing or rapid cooling. The steel after heat treatment therefore does not have long range residual stresses. It is very cheap to produce and has uniform properties in very large sections.

The reason for the high strength is well understood from the scale of the microstructure and the details of the compositions and fractions of the phases. However, the stress versus strain behavior is fascinating in many respects. Virtually all of the elongation is uniform, with hardly any necking. Indeed, the broken halves of each tensile specimen could be neatly fitted together. It is not clear what determines the fracture strain. In effect, the hard bainite has achieved all of the essential objectives of structural nanomaterials that are the subject of so much research, in large dimensions.

The potential advantages of the mixed microstructure of bainitic ferrite and austenite can be listed as follows:

1. Cementite is responsible for initiating fracture in high strength steels. Its absence makes the microstructure more resistant to cleavage failure and void formation.
2. The bainitic ferrite is almost free of carbon.
3. The microstructure derives its strength from the ultrafine grain size of the ferrite plates, which are less than 1μm in thickness. It is the thickness of these plates that determines the mean free slip distance, so that the effective grain size is less than a micrometer. This cannot be achieved by any other commercially viable process. It should be borne in mind that grain refinement is the only method available for simultaneously improving the strength and toughness of steels.
4. The ductile films of austenite that are intimately dispersed between the plates of ferrite have a crack blunting effect. They further add to toughness by increasing the work of fracture as the austenite is induced to transform to martensite under the influence of the stress field of a propagating crack.
5. The diffusion of hydrogen in austenite is slower than in ferrite. The presence of austenite can, therefore, improve the stress corrosion resistance of the microstructure.
6. Steels with the bainitic ferrite and austenite microstructure can be obtained without the use of any expensive alloying additions. All that is required is that the silicon concentration should be large enough to suppress cementite.

High strength nanobainitic steels are not as popular as quenched and tempered martensitic steels, because, in spite of these appealing features, the bainitic ferrite/austenite microstructure does not always give the expected good combination of strength and toughness. This is because the relatively large "blocky" regions of austenite between the sheaves of bainite readily transform into high carbon martensite under the influence of stress. This untempered, hard martensite embrittles the steel. This high carbon austenite can be stabilized by adding manganese and/or nickel so that the martensitic transformation ceases. Thus, a Fe–4Ni–2Si–0.4C% (3.69Ni, 3.85Si at %) alloy has been developed.

7.3 Metallurgy of High Strength Steels

High strength steels (HSS) are complex, sophisticated materials, with carefully selected chemical compositions and multiphase microstructures resulting from precisely controlled heating and cooling processes. Various strengthening mechanisms are employed to achieve a range of strength, ductility, toughness, and fatigue properties. These steels are not the mild steels of yesterday; rather they are uniquely lightweight and engineered to meet the challenges of today's vehicles for stringent safety regulations, reduced emissions, and solid performance, at affordable cost. It is a continuous challenge for the steel industry to develop steel grades that combine high strength with an acceptable ductility.

HSS are produced by controlling the chemistry and cooling rate from the austenite or austenite plus ferrite phase, either on the run out table of the hot mill (for hot rolled products) or in the cooling section of the continuous annealing furnace (continuously annealed or hot dip coated products). Research has provided chemical and processing combinations that have created many additional grades and improved properties within each type of AHSS.

Conventional low to high strength steels (mild, interstitial free, bake hardenable, and high strength low alloy steels) have simpler structures. AHSS are different from the conventional HSS because they were developed for increased strength and ductility for enhanced formability. The principal difference between conventional HSS and AHSS is their microstructure. Conventional HSS are single phase ferritic steels with a potential for some pearlite in C-Mn steels. AHSS are primarily steels with a microstructure containing a phase other than ferrite, pearlite, or cementite – for example, martensite, bainite, austenite, and/or retained austenite in quantities sufficient to produce unique mechanical properties. Some types of AHSS have a higher strain hardening capacity resulting in a strength ductility balance superior to conventional steels.

In martensitic steels (MS), nearly all austenite is converted to martensite. The resulting martensitic matrix contains a small amount of very fine ferrite and/or bainite phases. This structure typically forms during a swift quench following hot rolling, annealing, or a post forming heat treatment. Increasing the carbon content increases strength and hardness. The resulting structure is mostly lath (platelike) martensite. Careful combinations of silicon, chromium, manganese, boron, nickel, molybdenum, and/or vanadium can increase hardenability. The resulting martensitic steel is best known for its extremely high strength; UTS from 900 to 1700 MPa have been obtained. MS has relatively low elongation, but post quench tempering can improve ductility, allowing for adequate formability considering its extreme strength. Often used where high strength is critical, MS steel is typically roll formed and may be bake hardened and electro-galvanized for applications requiring corrosion resistance, but heat treating decreases its strength.

Because MS steel has such high strength to weight ratio, it is weight and cost-effective. Achieving the combination of both high strength and high ductility in the same steel has long been elusive metallurgically. Existing advanced high strength steels require that customers make a tradeoff between high strength parts that are designed using more limited geometries or using more expensive production methods such as hot stamping. The metallurgy and processing of HSS and AHSS grades are somewhat novel and unique compared to conventional steels.

Table 7.3 lists few most commonly used high strength and advanced high strength steels. Unlike low carbon steels, each high strength steel grade is identified by metallurgical type, minimum yield strength (in MPa), and minimum tensile strength (in MPa). As an example, DP 500/800 means a dual phase steel with 500 MPa minimum yield strength and 800 MPa minimum ultimate tensile strength.

TABLE 7.3

Mechanical properties most common high strength and advanced high strength steels.

Steel grade	YS, MPa	UTS, MPa	Elongation, %
BH 210/340	210	340	34–39
BH 260/370	260	370	29–34
DP 280/600	280	600	30–34
IF 300/420	300	420	29–36
DP 300/500	300	500	30–34
HSLA 350/450	350	450	23–27
DP 350/600	350	600	24–30
DP 400/700	400	700	19–25
TRIP 450/800	450	800	26–32
DP 500/800	500	800	14–20
CP 700/800	700	800	10–15
DP 700/1000	700	1000	12–17
Mart 950/1200	950	1200	5–7
Mart 1250/1520	1250	1520	4–6

7.4 Applications Particularly for Automotives

The major application of high strength steels being framed by automotive manufacturers is in automotives. The family of high strength steels (HSS) continues to evolve and grow in application, particularly in the automotive industry. New steel types are already being used to improve the performance of vehicles on the road, and emerging grades will be increasingly employed. By 1975, the average vehicle contained 3.6% medium and high strength steels for a total vehicle content of 61%, mostly mild steel. In the 1980s, the use of interstitial free (IF) and galvanized steels grew for complex parts. By 2007, the average vehicle contained 11.6% medium and high strength steels, for a total of 57%. The use of HSS in automobiles is quickly expanding with more research.

High strength low alloy (HSLA) steels, which had been used for major construction projects such as the Alaska Arctic Line Pipe Project in the 1970s, were increasingly developed and selected for automotive applications through the 1990s for their consistent strength, toughness, weldability, and low cost. Many automotive ancillary parts, body structure, suspension and chassis components, as well as wheels, are made of HSLA steel. Bake hardening (BH) steels are most commonly used for making closure panels like door outers, hoods, and decklids, as it can be hardened during painting cycle itself and have high dent resistance.

Dual phase (DP) steels are excellent in the crash zones of the car for their high energy absorption. These steels are used for automotive body panels, wheels, bumpers, crash boxes, support components, A, B, C, and D pillars in cars and box girders for chassis as they can be engineered or tailored to provide excellent formability for manufacturing complex parts for automobiles. DP is increasingly used by automakers in current car models. For example, in the 2011 Chevrolet Volt, the overall upper body structure is 6% DP by mass, and the lower structure is 15%, including such parts as the reinforcement for the rocker outer panel.

Transformation induced plasticity (TRIP) steels are some of the newest in development, but steel companies are quickly offering a greater variety of TRIP steels for automotive applications. They boast of its wide applicability, especially in complicated parts, and its high potential for mass savings. It can now be obtained in a variety of grades. Automotive applications of TRIP include body structure and ancillary parts. With high energy absorption and strengthening under strain, it is often selected for components that require high crash energy management, such as cross members, longitudinal beams, A and B pillar reinforcements, sills, and bumper reinforcements. For example, in 2007 Honda introduced the use of TRIP in the frame and side structure of the MDX, RDX, and CRV.

Patented steel wires are used in new generation tyres for heavy duty trucks and personal vehicles. Martensitic steels (MS) often used for body structures, ancillary parts, and tubular structures due to its high strength to weight ratio and cost effectiveness. MS grades are recommended for bumper reinforcement and door intrusion beams, rocker panel inners and reinforcements, side sill and belt line reinforcements, springs, and clips. For example, in the 2007 Honda Acura MDX, the rear martensitic bumper beam assists in rear crash protection.

In the past several years, automakers have increasingly incorporated HSS and AHSS into vehicles, especially structural and safety components. Research at Ducker Worldwide predicts that 50% of the average body in white (BIW) will be converted to HSS in this decade. Proposed regulations for 2017–2025 will impose much stricter Corporate Average Fuel Economy (CAFE) standards and, if passed, would potentially boost HSS use significantly. Current research aims to continue to expand the broad spectrum of HSS. One area of particular interest is the "third generation steels" to bridge the gap of strength-elongation balance between conventional and HSS steels and the austenitic-based steels. Some materials under development include nanosteels, bainitic steels, and so on.

8

High Alloy Steels

8.0 Introduction

The steels with more than 10% alloying elements are called high alloy steels. The most commonly used high alloy steels are maraging steels, stainless steels, and tool steels. This chapter describes these steels.

8.1 Maraging Steels

Conventionally, steels with yield strength higher than 700 MPa are called ultra-high strength steels. This level of strength can be achieved in normal low alloy steels with high carbon and alloy contents as discussed in previous chapters. However these steels will have problems such as:

(i) The strength depends mainly on carbon content. As carbon content increases weldability, machinability and formability are decreased
(ii) Special controls are needed to avoid decarburization
(iii) Distortion due to quenching
(iv) Will require high austenitizing temperature and
(v) Hydrogen embrittlement is a serious problem.

Many components in aerospace and space vehicles need steels with strength levels more than 1500 MPa and also without the problems mentioned above and this resulted in the discovery of MARtensite AGING steels called maraging steel, classified under ultra-high strength steels.

The development of the nickel maraging steels began in Inco research laboratories in the late 1950s and was based on the concept of using substitutional elements to produce age hardening in a low carbon nickel-iron martensitic matrix. Balanced additions of cobalt and molybdenum to iron nickel martensite gave a combined age hardening effect appreciably greater than the additive effects of these elements separately. The iron-nickel-

FIGURE 8.1
Microstructure of maraging steels

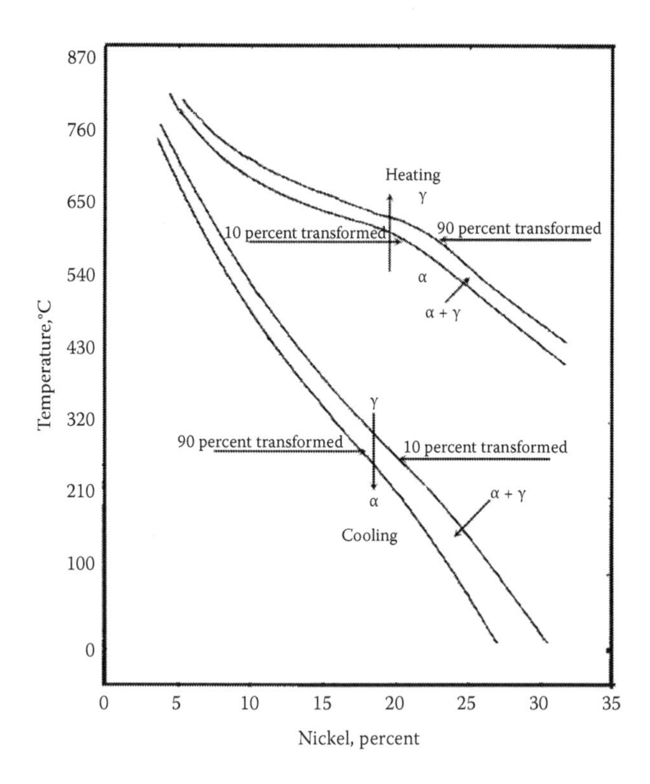

FIGURE 8.2
Iron-nickel phase diagram showing hysteresis loop

cobalt-molybdenum matrix was amenable to supplemental age hardening by small addition of titanium and aluminium. This resulted in the development of nickel-cobalt-molybdenum family of maraging steels. First commercial heat of maraging steel was made in December 1960.

8.1.1 Metallurgy of Maraging Steels

These steels have low carbon (less than 0.03%) and 20–25% nickel and other additions such as titanium, aluminium, chromium, cobalt, and so on. The iron nickel martensite (Figure 8.1) formed is very ductile due to the low carbon content and tempering is not required. The transformation temperature of austenite to martensite (α) upon heating and cooling has wide difference (hysteresis) as shown in Figure 8.2. So on reheating the iron nickel martensite, the martensite to austenite transformation will not occur at the same temperature at which austenite to martensite transformation occurred. This temperature difference, called as hysteresis, facilitates for precipitation hardening (aging) and thus the maraging steel is strengthened by both martensite and precipitates resulting in ultra-high strength.

If nickel content is more than 20%, austenite is retained upon cooling and martensite will not form. These type of steels can also be aged, that are called as ausaged steels. However, the strength of the ausaged steels are less than that of maraging steels and applications of ausaged steels are limited.

8.1.2 Effect of Composition

8.1.2.1 Cobalt and Molybdenum

Cobalt strengthens the matrix by solid solution strengthening while molybdenum dissolves in the matrix as well as forms carbides. The increase in strength of maraging steels is much higher if both cobalt and molybdenum are added together rather than individually as shown in Figure 8.3.

The reason for the synergic effect is as follows: Cobalt forms solid solution with iron and nickel and does not form either intermetallics or carbides to respond for aging. On the other hand, molybdenum forms carbides as well as strengthens the matrix by solid solution strengthening. When molybdenum alone is added in maraging steels, to some extent it dissolves in austenite and not available for forming carbide/intermetallic precipitates. Interestingly, addition of cobalt reduces the solubility of molybdenum in austenite and releases molybdenum for precipitation. The addition of molybdenum and cobalt increases notch tensile strength, ensures high value of reduction in area, and also reduces the tendency to stress corrosion cracking.

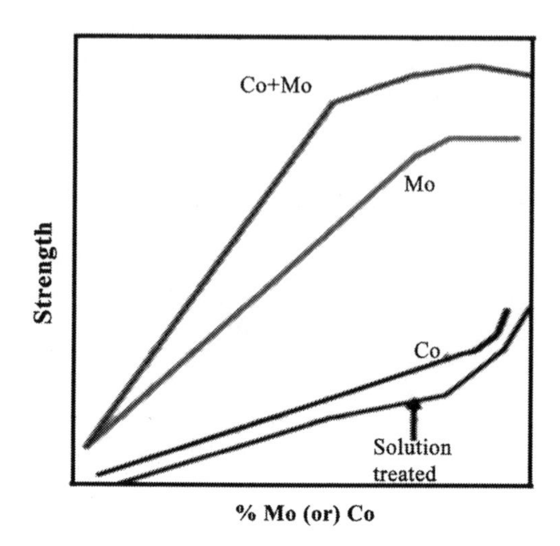

FIGURE 8.3
Synergic effect of cobalt and molybdenum in maraging steels

8.1.2.2 Titanium

Titanium is the main precipitation hardener in maraging steels. The addition of titanium between 0.6–0.7% does not affect notch tensile strength (NTS) and melting characteristics. However, more than 0.7% titanium affects air melting characteristics. Titanium also reduces carbon and nitrogen by forming TiC or TiN from martensite making martensite more ductile and tough.

8.1.2.3 Silicon and Manganese

Both silicon and manganese are detrimental elements in maraging steels since both decrease NTS. In many grades of steels nickel is replaced with manganese to stabilize the austenite but it is not used in maraging steels because it has a negative effect on NTS.

8.1.2.4 Carbon

The carbon content is restricted to a maximum of 0.03% as more than that reduces NTS although it is an economical hardener in steels.

8.1.2.5 Aluminium

Aluminium is not added intentionally as alloying element but up to 0.1% is allowable as it is used as deoxidizer during steel melting. Higher than 0.1% of aluminium decreases toughness.

8.1.2.6 Calcium

Maximum allowable limit is 0.02% as an impurity.

8.1.2.7 Boron and Zirconium

Boron and zirconium are added about 0.003% and 0.01%, respectively, as both retard grain boundary precipitation and thus increase toughness and reduce the tendency of stress corrosion cracking.

8.1.2.8 Chromium

Chromium is added in maraging steels where oxidation resistance is required. Maraging steels alloyed with chromium have substantially greater resistance to oxidation in air at 540°C than a 5% chromium tool steel.

8.1.3 Types and Composition of Maraging Steels

Maraging steels are basically iron nickel alloys with titanium. The other alloying elements such as cobalt, molybdenum, and chromium are added to improve various properties. The types are (i) Ni-Ti, (ii) Ni-Co-Mo-Ti-Al, (iii) Ni-Cr, and (iv) low alloy maraging steels. Like low alloy steels, there is no standardized designation. The numbers ascribed to the various grades of maraging steels are mostly based on proof stresses in SI units. For example, 18Ni1400 indicates the grade has 18% nickel and yield strength is 1400 MPa. It varies from country to country: for example, in the United States, the numbering refers to the nominal 0.2% proof stress values in kilo pounds per square inch (18Ni 250). Many grades are also numbered by the manufacturers. Table 8.1 shows composition and yield stress of the most commonly used maraging steels.

8.1.4 Mechanical Properties

Maraging steels have ultra-high strength combined with high ductility and high fracture toughness. Further, the strength of the maraging steels can be increased by cold working and subsequent aging. This type of process is called marforming. The maraging steels have good impact, fatigue, and fracture toughness. The impact fatigue limit of maraging steels ranges from 400 to 500 MPa.

The major requirements for aerospace landing gears and rocket cases are high fatigue strength to density and high tensile strength to density ratios and these are well satisfied by maraging steels. Table 8.2 shows the fatigue strength to density and tensile strength to density ratios of maraging steels along with other candidate materials.

TABLE 8.1

Composition and its yield stress of the most commonly used maraging steels

Grade	Proof stress, MPa	Nominal Composition, wt %						
		Ni	Co	Mo	Ti	Cr	Al	Others
Ni-Ti type								
25 Ni	1900	25	–	–	1.5	–	–	–
20 Ni	1900	20	–	–	1.5	–	–	–
Ni-Co-Mo-Ti-Al type								
18Ni1400	1400	18	8.5	3	0.2	–	0.1	–
18Ni1700	1700	18	8	5	0.4	–	0.1	–
18Ni1900	1900	18	9	5	0.6	–	0.1	–
18Ni2400	2400	17.5	12.5	3.75	1.8	–	0.15	–
17Ni1600 (cast)	1600	17	10	4.6	0.3	–	0.05	–
Ni-Cr type								
12–5–3	1270	12		3	0.2	3	0.3	–
IN 733	1562	10		–	0.3	12	0.7	–
IN 736	1270	10		2	0.2	10	0.3	–
IN 833	920	7		–	–	12	–	1 Si
Low alloy maraging steels								
IN 863	994	3	–	–	–	3	–	0.5 Si
IN 335	994	3	–	0.5	–	3	–	0.5 Si
IN 787	631	0.8	–	0.2	–	0.6	–	0.3 Si, 1.1Cu, 0.03Nb
IN 866	631	1	–	0.2	–	0.5	–	0.4 Si, 1.3 Cu, 1.8 Mn

TABLE 8.2

Fatigue strength to density and tensile strength to density ratios of candidate materials

Alloy	Fatigue strength/density	Tensile strength/density
Al 2024	30	68
Beta Ti alloy	32	120
8620 steels	21	57
Maraging steels	25	99

Though the best material is beta titanium alloys, it is not economical for commercial vehicles due to its high processing cost. The next best material is Al 2024 but the cross section area required to withstand the fatigue load results in bulky component and moreover, the tensile strength to density ratio is less. Hence the best alternative is maraging steels for aircraft landing gears.

The wear resistance of maraging steels are inferior to hardened/case-hardened steels in low stress abrasion conditions. However, case-nitrided maraging steels are better for high stress abrasion conditions because of its hard case and soft core. For gouging wear applications, maraging steels are better than austenitic stainless steels and Hadfield manganese steels due to its high toughness.

The fatigue strength of maraging steels varies between 0.41 and 0.53 of its tensile strength like conventional steels. However, maraging steels does not show any definite endurance limit like ferrous alloys and titanium.

8.1.5 Effect of Solutionising and Aging Temperature on Mechanical Properties

The achievable strength of the maraging steel with respect to the solutionizing temperature is schematically shown in Figure 8.4. The trend is more or less similar for most of the grades and types of maraging steels. As the solutionizing temperature increases, the tensile strength increases till a critical temperature above which there is drop in strength. Below the critical temperature, the precipitation strengtheners such as carbides, nitrides, and carbonitrides of titanium do not completely dissolve in the matrix resulting in coarse carbides and inhomogeneous distribution of the precipitates. Above the critical temperature, it is observed that the austenite is not converted to martensite completely leaving some amount of retained austenite that reduces the strength of the maraging steel. The critical temperature is about 760°C that may vary slightly depending on the alloying additions. However, the properties drastically come down if heated above 950°C due to uncontrollable grain growth. Care has to be taken in this regard.

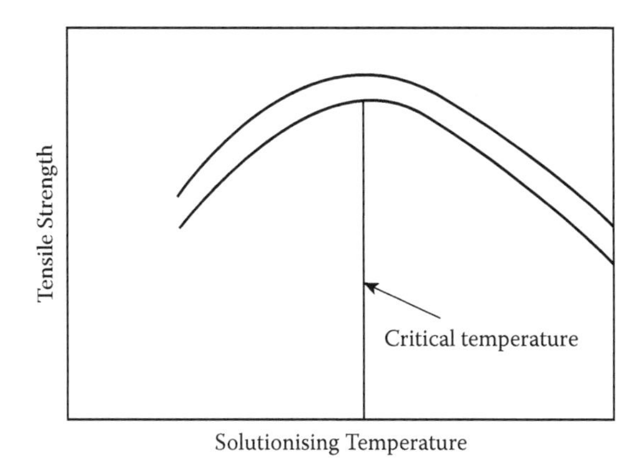

FIGURE 8.4
Effect of solutionizing temperature on mechanical properties

Similar to solutionizing temperature, aging temperature has significant role in achieving the best properties. Figure 8.5 schematically represents the effect of aging temperature on the mechanical properties of maraging steel. Most grades of maraging steels have the same trend and may slightly change based on the composition and type of maraging steel grades. In maraging steel, it is interesting to note that slight under aging or over aging does not affect the mechanical properties to a great extent.

8.1.6 Corrosion Behavior of Maraging Steels

In normal atmospheric condition, maraging steels corrode uniformly like normal steels and are covered with rust. However, the pits are smaller than conventional steels and corrosion rate is about half of the conventional steels. In marine atmosphere, the steels and maraging steels have the same corrosion rate but after about six months the corrosion rate of maraging steel reduces significantly. In general, maraging steels need to be protected under corrosive environments like other conventional steels. Maraging steels can be electroplated with chromium, nickel or cadmium where even slight corrosion is not tolerable. However, hydrogen that forms during plating operations may be absorbed by the steel causing embrittlement and cause delayed cracking (or) hydrogen induced cracking as shown in Figure 8.6. The effect due to absorbed hydrogen can be eliminated by baking for 24 hours in the temperature range of 200–320°C.

Maraging steels are prone to stress corrosion cracking (SCC) under aqueous environments with or without sodium chloride. However, maraging steels are better than other high strength steels offering better fracture

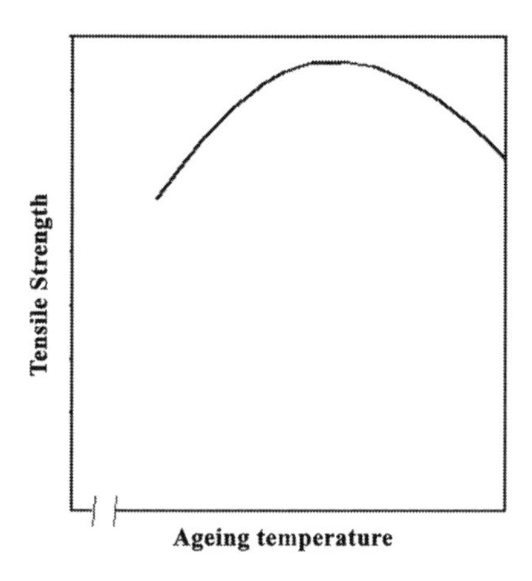

FIGURE 8.5
Schematic representation of effect of aging temperature on mechanical properties

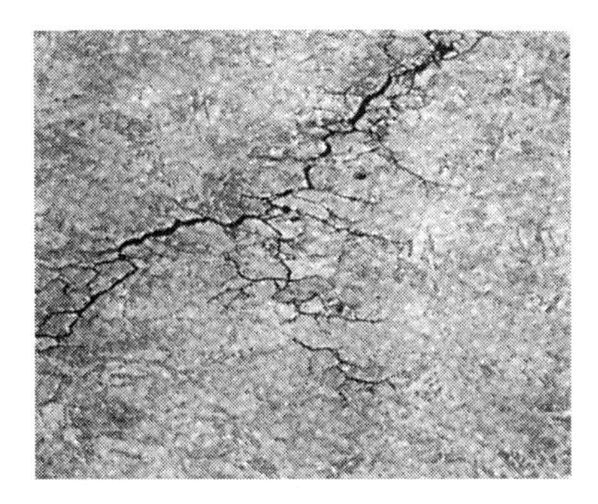

FIGURE 8.6
Delayed cracking due to hydrogen embrittlement in maraging steels

toughness values in SCC conditions i.e. can tolerate larger cracks without crack propagation under given static stress. SCC mainly depends on solutionizing temperature. Above 760°C, as the temperature increases, the time for cracking is reduced due to (i) grain growth and (ii) precipitation of TiC in the grain boundaries. This can be eliminated by suitable change in the composition so that the precipitation of TiC at grain boundaries is avoided.

8.1.7 Advantages of Maraging Steels

1. Ultra high strength with good ductility.
2. Insensitive to notches and the notch tensile strength ratio is close to 1.
3. Less prone to cracking and propagation of cracks due to high fracture toughness approx. 5500 $MPa/mm^{0.5}$
4. Tempering does not occur upon heating and maintains strength up to 350°C
5. Air cooling is sufficient to form martensite (good hardenability) and for precipitation hardening hence distortion due to cooling is less.
6. No decarburization problem due to ultra-low carbon (0.003%) and hence no need for special atmospheres for heat treatment that reduces furnace and handling cost.
7. Dimensional changes are minimum due to martensitic transformation because of very low carbon content as shown here:

   c = 2.861+ 0.116% C = 2.861348 = 0.012%

   a = 2.861–0.013% C = 2.860961 = -0.00136%

8. Have good machinability, weldability, formability.
9. Can be surface hardened or electroplated.
10. Corrosion resistance is much better than many high strength steels.

8.1.8 Applications in Aerospace

Because of high strength to weight ratio with good fracture toughness, good fabrication properties and simple heat treatment, maraging steels are widely used in critical areas such as aerospace – rocket motor cases, pressure hull in hydrospace, aircraft forgings, solid propellant missile cases, jet engine starter impellers, aircraft arrestor hooks, torque transmission shafts, aircraft ejector release units, landing gears, and so on. It is also used in structural engineering and ordnance – lightweight portable military bridges, ordnance components, and fasteners. Maraging steels are used for tooling and machinery – punches and die bolsters for cold forging, extrusion press rams and mandrels, extrusion dies for aluminium, cold reducing mandrels in tube production, die casting dies for aluminium and zinc base alloy, machine components like gears, index plates, lead screws, and so on.

Because the physical metallurgy and properties of maraging steels are unique, they have many commercial advantages and from the first day of production to till date the applications for maraging steels have steadily grown and widening. However, in automotive industries, maraging steels are not used due to its high cost; however, low alloy maraging steels may substitute in future some of the components of automobiles.

8.2 Stainless Steels

Stainless steels are iron-based alloys containing sufficient amount of chromium to form a chromium oxide layer (Cr_2O_3) of 1–3 nm thick. The chromium oxide layer is adherent to base material and self-healing. Thus, the steel achieves stainless character and hence got its generic name as stainless steel. It is also known as inox steel or inox, derived from the French word "inoxydable." The initial work started in England and Germany in 1910 and commercial production of stainless steels started in 1920 in the United States. Production of precipitation hardenable stainless steel started in 1945.

Stainless steels require a minimum of 10.5% of chromium to form the adherent self- healing chromium oxide layer to achieve its stainless properties. However, practically more chromium is added since chromium forms carbides first and chromium oxide is formed later. Sufficient amount of chromium is added to form the chromium oxide layer after forming the carbides.

Stainless steel is used for many critical applications and alloying elements such as nickel, molybdenum, copper, titanium, niobium, aluminium, and so on are added to improve specific properties demanded for specific applications.

8.2.1 Effect of Alloying on Structure and Properties

8.2.1.1 Chromium

It is the major alloying element in stainless steels. Minimum of 10.5% chromium is required for the formation of the protective layer of chromium oxide on the steel surface. The strength of this protective (passive) layer increases with increasing chromium content. Apart from this, chromium prompts the formation of ferrite within the alloy structure being a ferrite stabilizer. It also forms carbides and strengthens the steels.

8.2.1.2 Nickel

Nickel improves general corrosion resistance and prompts the formation of austenite. Stainless steels with 8–9% nickel have a fully austenitic structure. Increasing nickel content beyond 8–9% further improves both corrosion resistance (especially in acid environments) and workability.

8.2.1.3 Molybdenum and Tungsten

Molybdenum increases resistance to both local (pitting, crevice corrosion, etc.) and general corrosion. Molybdenum and tungsten are ferrite stabilizers and when used in austenitic alloys, must be balanced with austenite stabilizers in order to maintain the austenitic structure. Molybdenum is added to martensitic stainless steels to improve high temperature strength.

8.2.1.4 Nitrogen

Nitrogen increases strength and enhances resistance to localized corrosion. It is an austenite former.

8.2.1.5 Copper

Copper increases general corrosion resistance to acids and reduces the rate of work hardening.

8.2.1.6 Carbon

Carbon enhances strength (especially, in hardenable martensitic stainless steels), but may have an adverse effect on corrosion resistance by the formation of chromium carbides. It is an austenite stabilizer.

8.2.1.7 Titanium, Niobium and Zirconium

Where it is not desirable or, indeed, not possible to control carbon at a low level, titanium or niobium may be used to stabilize stainless steel against intergranular corrosion. As titanium (niobium and zirconium) have greater affinity for carbon than chromium, titanium (niobium and zirconium) carbides are formed in preference to chromium carbide and thus localized depletion of chromium is prevented. These elements are ferrite stabilizers.

8.2.1.8 Sulfur

Sulfur is added to improve the machinability of stainless steels. Sulfur bearing stainless steels exhibit reduced corrosion resistance.

8.2.1.9 Cerium

Cerium, a rare earth metal, improves the strength and adhesion of the oxide film at high temperatures.

8.2.1.10 Manganese

Manganese is an austenite former, that increases the solubility of nitrogen in the steel and may be used to replace nickel in nitrogen-bearing grades.

8.2.1.11 Silicon

Silicon improves resistance to oxidation and is also used in special stainless steels exposed to highly concentrated sulfuric and nitric acids. Silicon is a ferrite stabilizer.

8.2.2 Classification

Based on the room temperature microstructure, the stainless steels are classified as;

1. Martensitic stainless steels
2. Ferritic stainless steels
3. Austenitic stainless steels
4. Duplex (ferritic-austenitic) stainless steels
5. Precipitation hardenable stainless steels

The AISI has classified the wrought and cast stainless steels with three digit numbers. The 400 series is assigned for martensitic and ferritic stainless steels. Austenitic stainless steels are specified with 300 series and 200 series. In 200 series austenitic stainless steels, nickel is replaced by manganese. The duplex stainless steels and precipitation hardenable stainless steels are mostly designated by the manufacturers.

8.2.2.1 Martensitic Stainless Steel

Martensitic stainless steels contain chromium about 10.5 to 18% and carbon of maximum 1.2%. Chromium is a ferrite stabilizer and carbon is austenite stabilizer. These elements should be balanced in martesitic stainless steel in such a way that on heating it should form austenite and only then upon quenching it will form BCT martensite at room temperature (Figure 8.7). If it is not balanced, chromium will stabilize ferrite even at high temperature and austenite bay will be completely vanished as shown in Figure 8.8. And also it should be ensured that chromium availability after forming carbide should be at least 10.5% to form adherent self-healing protective layer to achieve stainless properties. Essentially, the chromium content should be higher if the carbon amount is increased.

Martensitic stainless steels are heat treatable and have more wear resistance but corrosion resistance is less than the other types. Chemical composition and mechanical properties of most commonly used martensitic grade stainless steels are given in Table 8.3.

Oil quenching these alloys from temperatures between 982–1066°C produces the highest strength and/or wear resistance as well as corrosion resistance. The as quenched structure of fresh martensite must be tempered to restore some ductility. The values given in Table 8.3 are for samples tempered at temperatures from 204°C to 649°C for two hours.

FIGURE 8.7
Microstructure of martensitic stainless steel

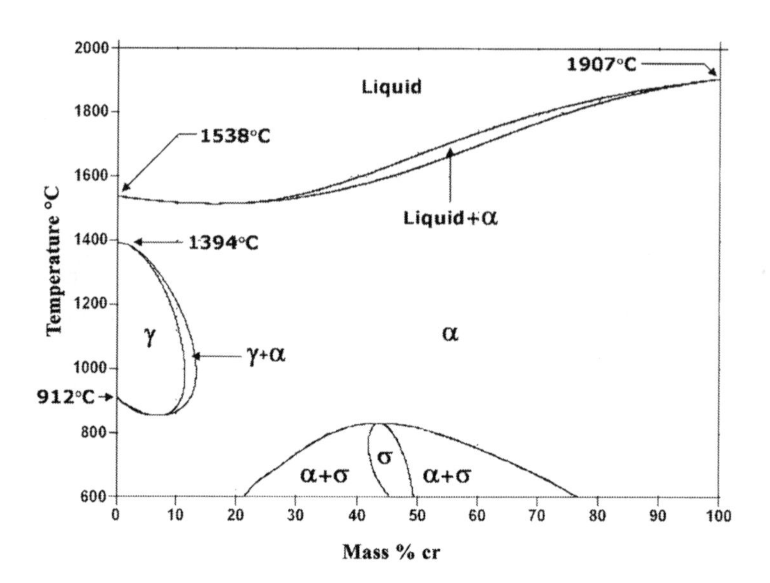

FIGURE 8.8
Iron-chromium phase diagram

8.2.2.2 Ferritic Stainless Steels

Ferritic stainless steels contain 10–30% chromium. At room temperature, it has BCC ferrite (Figure 8.9). Its corrosion resistance is better than martensitic stainless steel. It gives very good surface finish and hence used for architectural purpose. Table 8.4 shows the composition and mechanical properties of the most important ferritic stainless steels.

Ferritic stainless steels are used for their anticorrosion properties rather than for their mechanical properties (strength, ductility, or toughness). Initially, ferritic stainless steels were used in non-welded constructions (riveted and bolted assemblies, etc.).

The ferritic stainless steels are generally classified as (i) first-generation, (ii) second-generation, and (iii) third-generation. The first generation includes mainly the following grades: 430, 434, 436, 442 and 446. In addition to their high chromium content, those containing 0.12% to 0.20% of carbon are vulnerable to a drop in ductility and to corrosion due to the formation of martensite and intermetallic precipitates in the ferritic phase in the as welded condition. The first-generation steels were supposed to be nonweldable, particularly for thicker sections that have a high inclination to cold cracking and exhibit a drastic drop in ductility in the as welded condition. The low weldability of these alloys is mainly due to the high carbon and chromium contents. The difficulty in obtaining alloys with a low level of impurities and undesirable residual elements sulfur,

TABLE 8.3

Chemical composition and mechanical properties of most commonly used martensitic grade stainless steels

Type	Elements in wt%						Mechanical Properties				
	C	Mn	Si	Cr	Mo	Ni	HRB	YS (MPa)	TS (MPa)	% elongation	Hardening response (HRc)
410*	0.15 max	1 max	1 max	11.5–13.5	–	0.5 max	82	290	510	34	38–45
410 HC**	0.21	1 max	1 max	12.5	–	0.5 max	83	310	538	30	45–52
420	0.15–0.4	1 max	1 max	12–14	–	0.5 max	87	310	586	29	53–57
425 Mod	0.5–0.55	1 max	1 max	13–14	0.8–1.2	0.5 max	93 89*	379 310*	648 593*	24 25*	57–60
440 A	0.6–0.75	1 max	1 max	16–18	0.75 max	0.5 max	95	427	717	20	57–60
420HC**	0.44	1 max	1 max	13	–	0.5 max	88	310	600	28	56–59

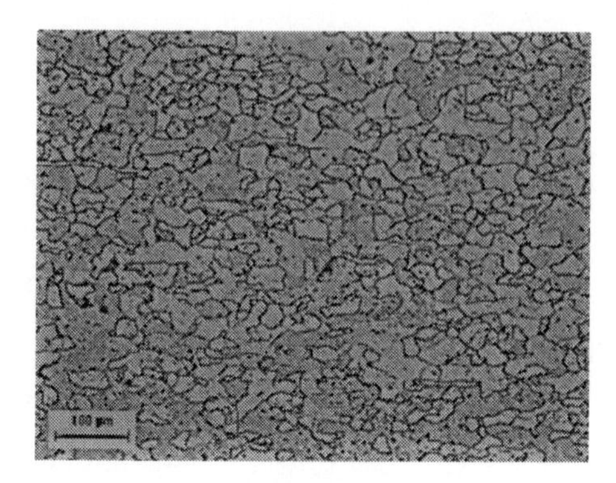

FIGURE 8.9
Microstructure of ferritic stainless steel

phosphorus, carbon, and so on have restricted the weldability of ferritic stainless steels.

The invention of the AOD (argon oxygen decarburization) process in the 1960s has enabled the development of high purity iron-chromium alloys containing negligible traces of carbon. Since then, the development of other generation of ferritic stainless steels had been started. The second generation includes low carbon alloys developed from the modification of first-generation grades. They contain other ferrite forming elements like titanium, niobium, and aluminum. In addition to promoting the ferritic microstructure, aluminum also improves the oxidation resistance at high temperatures by stabilizing the chromium oxide layer. This generation includes 405, 409, 409Cb, 439, and so on grades, and the weldability of these alloys is noticeably better than the first generation grades.

The third generation is of recent alloys called superferritic. The forming process of these alloys experiences some difficulties related to the elevated chromium levels. These are used for applications subjected to severe corrosion conditions such as in chloride environments. The grades 444, 29-4, 29-4-2 are the most typical ones in this class. These are high purity alloys in which the amount of interstitial elements, carbon and nitrogen in particular, are reduced to a minimum in order to improve the ductility of the alloy.

However, various types of hot embrittlement (loss of ductility) are associated with ferritic stainless steels that should be taken care when used at high temperatures.

1. ***Embrittlement at 475°C***: This phenomenon is proportional to the chromium content and occurs in the temperature range of 425–550°C. It is the result of the decomposition of the ferrite into two phases: a chromium rich phase (α') and an iron rich phase (α). The more the

TABLE 8.4

Chemical composition and mechanical properties of important grades of ferritic stainless steels

	Chemical Composition (in wt%)						Mechanical Properties			
Type	**C**	**Cr**	**Ni**	**Mn**	**Si**	**Other**	**Yield strength, MPa**	**Tensile strength, MPa**	**Elongation (%)**	**Hardness (BHN)**
405	0.06	11.5–14.5	0.5	1	1	Al-0.1–0.3	221	414	20	180
409	0.03	10.5–11.7	–	1	1	Ti= 6 x %C	175	360	20	150
409 Cb	0.03	17.5–19.5	1	1	1	N= 0.03, Ti-0.1–0.5, Nb = 0.3+9X%C	324	496	30	147
430	0.12	14–18	0.5	1	1	–	242	414	20	200
434	0.05	17	–	0.4	0.25	Mo-0.9	340	490	25	180
436	0.05	17	0.5	0.4	0.25	Mo-0.8	340	520	30	180
439	0.03	17–19	0.5	1	1	N=0.03, Al=0.15, Ti=0.2+4(C+N)	207	414	22	180
442	0.12	20	–	-	-	–	311	518	20	200
444	0.025	17.5–19.5	1	1	1	Nb = 1, Ti=0.2+4(C+N)	275	415	20	217
446	0.2	23–27	0.5	1.5	1	N-0.25	311	518	20	200
29–4	0.01	28–30	0.15	0.3	0.2	Mo=3.5–4.2	450	600	23	200
29–4-2	0.01	28–30	2–2.5	0.3	0.2	Mo=3.5–4.2	415	550	20	230

alloy is rich in chromium, the faster will be the reaction. The consequences of this reaction consist essentially in a selective corrosion of the iron rich phase. In order to prevent this type of embrittlement, the service temperatures of the part must always be limited to 400°C max.

2. ***Precipitation of embrittling phases and intergranular corrosion Sigma Phase:***

   Sigma (σ) phase (a chromium-rich and very brittle phase) is the product of transformation of the ferritic delta phase occurring when the microstructure is subjected to a long exposure time in the temperature range of 540–870°C, as shown in Figure 8.10.

This new phase (σ) forms rapidly in high chromium grades and molybdenum also promotes its formation. The presence of this new phase alters both the corrosion resistance and ductility. Dissolving this phase requires a solutionizing heat treatment at temperatures above 900°C. However, ferritic stainless steels are limited to 400°C maximum and hence the possibility of formation of sigma phase is remote during service.

8.2.2.3 Austenitic Stainless Steels

Austenitic stainless steels have FCC austenite phase at room temperature and contain typically 16–25% chromium and 8–26% nickel. These also contain nitrogen in solution as austenite stabilizer. Nonavailability of nickel during 1950s due to civil war in Africa and Asia a new series called 200 series evolved that had manganese instead of nickel to stabilize austenite. Due to less cost and good strain hardening, the 200 series are widely used for noncritical applications. Austenitic stainless steels make up over 70% of total stainless steel production and is more expensive than

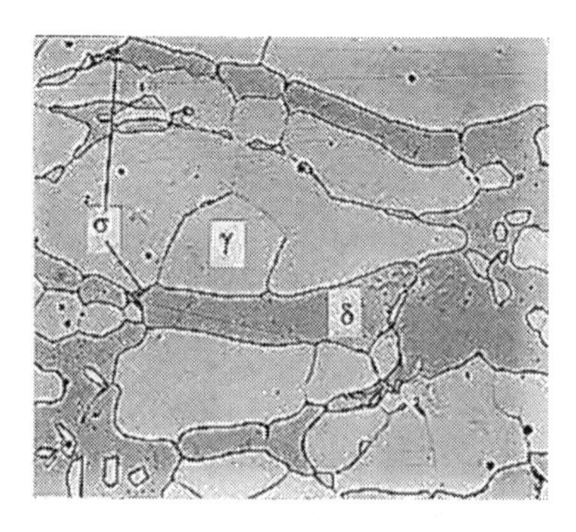

FIGURE 8.10
Microstructure showing formation of sigma (σ) phase

the other types. Type 304 is the most common type among all austenitic stainless steels, and the most weldable one and it can be welded by all fusion and resistance welding processes. Table 8.5 shows composition and mechanical properties of most commonly used austenitic stainless steels.

Austenitic stainless steels are nonmagnetic and are extremely formable and weldable. They can be successfully used from cryogenic temperatures to the red hot temperatures. Austenitic (FCC) structure is very tough and ductile down to absolute zero as they have no ductile to brittle temperature (DBTT) like ferritic structure. Austenite is a high temperature phase and more densely packed (FCC) and hence austenitic stainless steels do not lose their strength at elevated temperatures as rapidly as ferrite (conventional steels). So, austenitic stainless steel can be used for high temperature applications also. In addition, they are soft enough (i.e., with a yield strength about 200 MPa) to be easily formed by the same tools that work with carbon steel and they can also be made incredibly strong by cold work, up to yield strengths of over 2000 MPa. These properties make the austenitic steel a versatile material for varying applications. But for the cost of the nickel these alloys would be widely used.

8.2.2.4 Lean and Richer Austenitic Stainless Steels:

Lean alloys have less than 20% chromium and 14% nickel and constitute the largest portion of all stainless steel produced. These are principally 201, 301, and 304. The main difference among the lean austenitic alloys lies in their work hardening rate. The leaner the alloy, the lower the austenite stability. As unstable alloys are deformed, they transform from austenite to the much harder martensite. This increases the work hardening rate and enhances ductility.

Richer alloys, such as 305, have lowest work hardening rate. The austenite is more stable and will not transform to martensite on deformation like lean alloys. These grades are preferred for those applications where corrosion predominate.

8.2.2.5 Austenitic Stainless Steels for High Temperature Oxidation Resistance:

High temperature oxidation resistance can be enhanced by addition of silicon and rare earths. If the application requires also strength at high temperature, carbon, nitrogen, niobium, and molybdenum are added. 302B, 309, 310, 347, and various proprietary alloys are found in this group.

Chromium, molybdenum, nickel, and nitrogen alloyed austenitic stainless steel is used when corrosion resistance is the main objective. Alloys such as silicon, molybdenum, nitrogen and copper are added for resistance to specific environments. This group includes 316L, 317L, 904L, and many proprietary grades.

TABLE 8.5

Composition and mechanical properties of most commonly used austenitic stainless steels

	Chemical Composition (in wt%)						Mechanical properties			
Type	**C**	**Cr**	**Ni**	**Mn**	**Si**	**Other**	**Yield strength, MPa**	**Tensile strength, MPa**	**Elongation (%)**	**Hardness (BHN)**
201	0.15	16–18	3.5–5.5	5.5–7.5	1	0.25 N	276	793	40	210
202	0.15	17–19	4–6	7.5–10	1	0.25 N	276	690	40	210
301	0.15	16–18	6–8	2	1	–	242	690	50	180
302	0.15	17–19	8–10	2	1	–	207	552	50	180
302 B	0.15	17–19	8–10	2	2–3	–	205	515	40	180
304	0.08	19	10	2	1	–	207	518	40	180
304L	0.03	19	10	2	1	–	207	518	40	180
305	0.12	17–19	10–13	2	1	–	205	515	40	200
309	0.2	22–24	12–15	2	1	–	207	518	40	200
310	0.25	24–26	19–22	2	1	0.03N	247	571	50	–
316	0.08	16–18	10–14	2	1	Mo 2–3	207	518	40	200
316L	0.03	16–18	10–14	2	1	Mo 2–3	207	518	40	200
316LN	0.03	16–18	10–14	2	1	Mo 2–3 N-0.03	207	518	40	200
317	0.08	18–20	11–15	2	1	3–4 Mo	205	515	40	200
317L	0.03	18–20	11–15	2	1	3–4 Mo	240	585	55	205
321	0.08	18	10	2	1	0.4 Ti	242	587	55	205
347	0.08	18	11	2	1	0.8Nb	242	621	50	210
904L	0.02	19–23	23–28	2	1	4–5 Mo 1–2 Cu	220	490	35	180

The austenitic stainless steels that require enhanced machinability have a high content of controlled inclusions, sulphides, or oxy sulphides, to improve machinability at the expense of corrosion resistance. Carbon kept below 0.03% and designated as L grade is used when prolonged heating due to multipass welding of heavy section (greater than about 2 mm) or when welds requiring a postweld stress relief are anticipated.

Austenitic stainless steels are available as both wrought and cast alloys. The composition of wrought and cast alloys are almost the same except that the cast counterparts differ primarily in silicon content.

8.2.2.6 Austenite Stability

The formation of martensite at room temperature is thermodynamically possible in the case of austenitic stainless steels, but the driving force for its formation may be insufficient for it to form spontaneously. As it is known that martensite forms from austenite by diffusion- less shear mechanism, it can also occur if that shear is provided mechanically by external forces. This is called as strain induced martensite. The transformation temperature and the degree to which it occurs varies with composition of the alloy. The following equation relates the temperature at which 50% of the austenite transforms to martensite with 30% true strain (Md30).

$$Md30\ (^\circ C) = 551 - 462(\%C + \%N) - 9.2(\%Si) - 8.1(\%Mn) - 13.7(\%Cr) - 29(\%Ni + Cu) - 18.5(\%Mo) - 68(\%Nb) - 1.42\ (\text{grain size in microns} - 8)$$

All the alloying elements (ferrite and austenite stabilizers) decrease the Md30 temperature and delay martensite formation i.e. if austenitic stainless steel is deformed above Md30 temperature, the austenite will be stable and will not form any martensite and if the deformation temperature is lower the austenite transforms to martensite when the true strain is 30%. Figure 8.11 represents schematically the transformation of austenite to martensite at different strain and temperature.

This temperature is the common index of austenite stability. The lean alloys obviously have the Md30 temperature higher than room temperature and hence it forms martensite even during deforming at room temperature, that contributes for the high strain hardening. The actual Md30 temperature can be found out for each alloy based on the composition and thus formation of strain-induced martensite can be avoided. Because the carbon levels of austenitic stainless steels are always relatively low, strain induced martensite is self tempered and not brittle. Martensite thus formed is, of course, susceptible to hydrogen embrittlement. The strain induced martensite can be eliminated by heating the austenitic stainless steels to high temperatures (normally above 800°C) and subsequent quenching is called as quench annealing.

Care has to be taken while using the regression equation since this regression analysis was generated for homogeneous alloys. If alloys are inhomogeneous such as sensitized or solute segregated, due to welding,

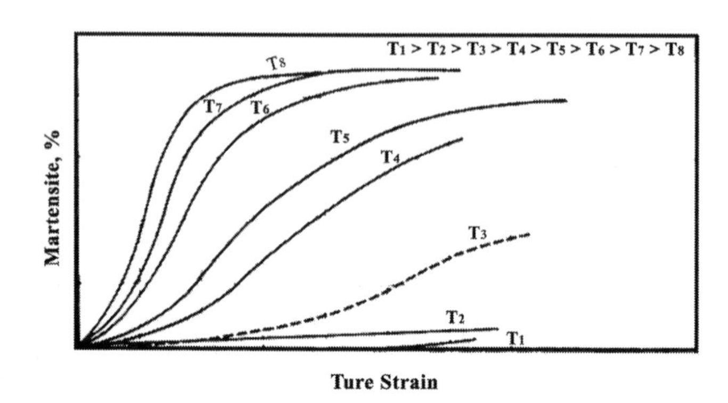

FIGURE 8.11
Variation of martensite formation with temperature and true strain (Schematic)

then the equation applies on a microscopic scale. Sensitized zones (i.e., the regions near grain boundaries where chromium carbides have precipitated) will have a much higher tendency to transform to martensite.

8.2.2.7 Mechanical Properties

The tensile properties in the annealed state related to composition as given by the empirical equation.

Yield strength (MPa) = 15.4 [4.4+23(%C)+32(%N)]+0.24 (%Cr)+0.94 (%Mo) + 1.3 (%Si) + 1.2(%V)+0.29(%W)+2.6(%Nb)+1.7(%Ti)+0.82(%Al)+0.16(% ferrite)+ 0.46 ($d^{-0.5}$)

Tensile strength (MPa) = 15.4 [29+35(%C)+55(%N)]+ 2.4(%Si)+ 0.11(%Ni)+ 1.2(%Mo)+ 5(%Nb)+3(%Ti)+1.2(Al%)+0.14(% ferrite)+0.82 ($d^{-0.5}$)

where percentage of elements refer to weight percentage and d is the grain diameter in millimeters. There are several empirical relations used to calculate the mechanical properties. In general as the alloying elements increase, the strength increases. The tensile strength relationship does not obey for lean alloys, such as 301, in which tensile strength increases with decreasing alloy content due transformation of austenite to martensite, that produces higher tensile strengths in austenitic stainless steels.

One important point that design engineers should take care is that austenitic stainless steels do not have a clear yield point but can begin to deform at 40% of the yield strength to some extent. As a rule of thumb, behavior at less than half the yield strength is considered fully elastic and stresses below two thirds of the yield strength produce negligible plastic deformation. This quasi-elastic behavior is due to many active slip systems in the FCC structure. The tensile properties of austenitic stainless steels with unstable austenite (lean alloys), are very strain rate dependent due to the influence of adiabatic heating during testing increasing the stability of the austenite. Austenitic stainless steels have exceptional toughness. The ambient temperature impact strength of austenitic stainless steels is quite high. Absence of a transition temperature (DBTT) make the austenitic stainless steel suitable for cryogenic applications.

8.2.2.8 **Precipitation of Carbides**

The solubility of carbon is very little at room temperature and even the 0.03% of L grade is mostly in a supersaturated solution. Because carbon has a great thermodynamic affinity for chromium and forms chromium carbides of $M_{23}C_6$ form. Formation of chromium carbides causes local chromium depletion adjacent to carbides such that the chromium level can become low enough not to be stainless resulting in much lower corrosion resistance than the surrounding area.

Interestingly in normal conditions, austenitic stainless steels are not having chromium carbides due to the slow diffusion of carbon and the even slower diffusion of chromium in austenite. However, on reheating and prolonged holding at high temperatures will precipitate chromium carbides and cause depletion of chromium near carbides resulting in localized corrosion is called "sensitization." In general at low temperatures, grain boundary diffusion is much more rapid than bulk diffusion, and grain boundaries provide excellent nucleation sites and hence precipitation of chromium carbides occurs along grain boundaries.

The temperature and incubation time for chromium carbide formation (time for sensitization) is depended on the amount of carbon in the austenitic stainless steels. For example, in austenitic stainless steel with 0.06% C (normal 304 grades), the chromium carbides start precipitating from 475 to 850°C and the time varies between 1 min to 100 h depending on the temperature which is called as sensitization zone where as for austenitic stainless steel with 0.02% C sensitization zone is between 475 to 580°C and time is above 100 h. Care has to be taken while using the austenitic stainless steel at high temperatures and also during welding by ensuring that the alloy is not cooled slowly in the sensitization temperature range. Even the austenitic stainless steels should be quenched from annealing temperature even for annealing is called as quench annealing.

Much longer heat treatment is required to eliminate these depleted zones by re homogenization of slowly diffusing chromium although only short time is required to form them. Alloying elements can have a major influence on carbide precipitation by their influence on the solubility of carbon in austenite. Molybdenum and nickel accelerate the precipitation by diminishing the solubility of carbon. Chromium and nitrogen increase the solubility of carbon and thus retard and diminish precipitation. Nitrogen is especially useful in this regard.

8.2.2.9 Stabilization

The composition of austenitic stainless steels are adjusted in such a way that the material will not be subjected to sensitization is called stabilized austenitic stainless steels or stabilization. The carbon is lowered to harmless levels, so that free carbon is not available to form carbides. These grades are mentioned as L grade. 304 L refers to the reduced carbon level of 304 grade. Normally L grade has carbon lower than 0.03%. The other way is to add nitrogen, that increases the solubility of carbon in austenite and are mentioned as LN grades, that is, 304 LN. It was found that adding more powerful carbide formers than chromium could preclude the precipitation of chromium carbides. Titanium and niobium are the most useful elements in this regard. Titanium above four times the carbon and niobium eight times the carbon avoid the formation of chromium carbide precipitation and sensitization.

8.2.2.10 Weaknesses

Austenitic stainless steels are less resistant to cyclic oxidation than ferritic grades because their greater thermal expansion coefficient tends to cause the protective oxide coating to spall. They can experience stress corrosion cracking (SCC) if used in an environment for which they have insufficient corrosion resistance. The fatigue endurance limit is only about 30% of the tensile strength (vs. ~50 to 60% for ferritic stainless steels). This, combined with their high thermal expansion coefficients, make them especially susceptible to thermal fatigue.

8.2.2.11 Duplex Stainless Steels

The duplex stainless steels have both ferrite and austenite at room temperature. The ferrite stabilizer chromium about 23–28% and austenite stabilizer nickel about 2.5–5% are added to achieve the duplex structure shown in Figure 8.12. The austenite islands (light) are embedded in a continuous ferrite matrix. The duplex structure contains about 45–65% austenite and the rest is ferrite. The amount of ferrite and austenite can be adjusted by balancing chromium and nickel contents.

Duplex stainless steels were created to combat corrosion problems caused by chloride bearing cooling waters and other aggressive chemical process fluids. The first-generation duplex stainless steels were developed more than 70 years ago in Sweden for use in the paper industry. The corrosion performance is evaluated by resistance to pitting and termed as pitting resistance equivalent, that depends on their alloy content.

Pitting resistance equivalent = Cr% + 3.3Mo% + 16N%

The term "Super Duplex" was first used in the 1980's to denote highly alloyed, high-performance duplex stainless steel with a pitting resistance equivalent of >40. With its high level of chromium, super duplex steel

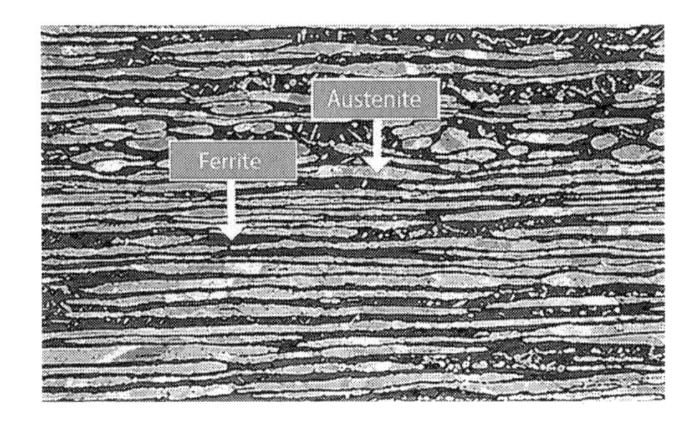

FIGURE 8.12
Microstructure of duplex stainless steel

provides outstanding resistance to acids, acid chlorides, caustic solutions and other environments in the chemical/petrochemical, pulp and paper industries, often replacing 300 series stainless steel, high nickel super austenitic steels and nickel-based alloys.

The chemical composition based on high contents of chromium, nickel and molybdenum improves intergranular and pitting corrosion resistance. Additions of nitrogen promote structural hardening by interstitial solid solution mechanism, that raises the yield strength and ultimate strength values without impairing toughness. Moreover, the two-phase microstructure guarantees higher resistance to pitting and stress corrosion cracking in comparison with conventional stainless steels.

Duplex stainless steels are graded for their corrosion performance depending on their alloy content.

- Lean Duplex such as 2304, that contains no deliberate molybdenum addition;
- 2205, has molybdenum, the work-horse grade accounting for more than 80% of duplex usage;
- 25 Cr duplex such as Alloy 255 and DP-3;
- Super-Duplex; with 25–26 Cr and increased molybdenum and nitrogen compared with 25 Cr grades, including grades such as 2507, Zeron 100, UR 52N+, and DP-3W.

Most of the duplex stainless steels are proprietary grades but few grades had been identified by UNS number. Chemical composition of most commonly used duplex stainless steels are given in Table 8.6.

The benefits of duplex stainless steels are as follows: (i) high strength, (ii) high resistance to pitting, crevice corrosion, stress corrosion cracking,

TABLE 8.6

Chemical composition (in wt%) of most commonly used duplex stainless steels

UNS number	Type	Chemical Composition, wt%							
		C	Mn	Si	Cr	Ni	Mo	N	Cu	Other
S31200	–	0.03	2	1	24–26	5.5–6.5	1.2–2	0.14–0.2	–	–
S31260	–	0.03	1	0.75	24–26	5.5–7.5	2.5–3.5	0.1–0.2	0.2–0.8	W-0.1–0.2
S31803	–	0.03	2	1	21–23	4.5–6.5	2.5–3.5	0.08–0.2	–	–
S32001	–	0.03	4–6	1	22–23	1–3	0.6	0.05–0.17	1	–
S32205	2205	0.03	2	1	19.5–21.5	4.5–6.5	3–3.5	0.14–0.2	–	–
S32304	2304	0.03	2.5	1	21.5–24.5	3–5.5	0.05–0.6	0.05–0.2	0.06–0.6	–
S32520	–	0.03	1.5	0.8	24–26	5.5–8	3–4	0.2–0.35	0.5–2	–
S32550	255	0.04	1.5	1	24–27	4.5–6.5	2.9–3.9	0.1–0.25	1.5–2.5	–
S32750	2507	0.03	1.2	0.8	24–26	6–8	3–5	0.24–0.32	0.5	–
S32760	–	0.03	1	1	24–26	6–8	3–4	0.2–0.3	0.5–1	–
S32900	329 d	0.06	1	0.75	23–28	2.5–5	1–2	–	–	–

corrosion fatigue and erosion, (iii) high thermal conductivity and low coefficient of thermal expansion, (iv) good workability and weldability, and (v) high energy absorption.

8.2.2.12 Precipitation Hardenable Stainless Steels

The precipitation hardenable steels as the name indicates are precipitation hardenable. Copper, molybdenum, and aluminium are added to strengthen by precipitation hardening the austenitic phase or martensite phase. The precipitation hardenable stainless steels are of three types: martensitic, semiaustenitic, and austenitic. Table 8.7 shows the composition of most commonly used precipitation hardenable stainless steels.

Martensitic group steels are solutionized at 1040°C and air cooled so that austenite is transformed to martensite. Aging in temperature range of 455–565°C causes precipitation effect. These grades can have tensile strength of 1345 MPa and yield strength of 1241 MPa with 13% elongation in the fully aged condition. These alloys have poor cold forming and shearing characteristics. The compositions of the semiaustenitic type precipitation hardenable stainless steels such as 17–7 PH and PH 15–7 Mo are adjusted in such a way that martensite start (Ms) temperature is well below room temperature. When these stainless steels are quenched from solutionizing temperature, the austenite is stable and martensite will not form. At this condition they are soft and ductile and can be easily formed. The formed parts are subjected to conditioning treatment, that is, they are heated to a temperature between 760 and 950°C, so that the carbon comes

TABLE 8.7

Composition of most commonly used precipitation hardenable stainless steels

Alloy	Composition in wt %							
	C	Cr	Ni	Mo	Al	Mn	Si	others
Martensitic								
PH 13–8 Mo	0.05	12.25–13.25	7.5–8.5	2–2.5	0.9–1.3	0.1	0.1	0.01N
15–5 PH	0.07	14–15.5	3.5–5.5	–	–	1	1	2.5–4.5 Cu, 0.14–0.45 Nb
17–4 PH	0.07	16.5	4	–	–	1	1	Cu-2.75
Stainless W	0.07	17	7	–	0.2	–	–	Ti- 0.7
Semiaustenitic								
17–7 PH	0.07	17	7	–	1.15	0.6	0.4	–
PH 15–7 Mo	0.09	15	7	2.5	1	1	1	–
AM 350	0.1	16.5	4.3	2.75	–	0.8	0.25	N- 0.1
AM 355	0.13	15.5	4.3	2.75	–	0.95	0.25	N- 0.1
Austenitic								
17–10 P	0.12	17	10	–	–	–	–	P- 0.25
HNM	0.3	18.5	9.5	–	–	3.5	–	P- 0.23

out from austenite and forms respective carbides and brings the Ms and Mf temperature above room temperature. Upon cooling, the austenite transforms in to martensite. Hence they are called as "semiaustenitic." If the conditioning temperature is at the lower side (around 760°C), Mf will be above room temperature but if the conditioning temperature is at the higher side (around 950°C), Mf temperature will be below zero and in such cases subzero cooling (or) refrigeration is required to complete the martensitic transformation. The martensite obtained due to high temperature conditioning is harder due to high carbon content. Alternatively, the austenitic to martensitic transformation can also be achieved by cold working. However in this, the subsequent aging is carried out between 455 and 565°C for 1 to 3 h. In case of austenitic precipitation hardenable stainless steels, the Ms temperature is well below zero and thus the transformation of austenite is not feasible. Aging is carried out in the austenitic matrix itself to get its optimal mechanical properties. The aging kinetics of the austenitic type is much slower as compared to semi austenitic or martensitic type. The interesting fact about the alloys AM 350 and 355 is that they do not respond to precipitation hardening but the heat treatment cycle is similar to that of semiaustenitic type and hence they are also classified under semi austenitic precipitation hardenable stainless steels. However, the martensite produced in AM 350 and 355 are harder than other alloys due to high carbon and thus the strength of these alloys are higher than other alloys.

Thus the required mechanical properties can be obtained in precipitation hardenable stainless steels by selection of proper heat treatment cycle rather than adjusting its composition.

8.2.3 Weldability of Stainless Steel

The unique properties of the stainless steels are derived from the addition of alloying elements, principally chromium and nickel, to steel. Typically, more than 10% chromium is required to produce a stainless iron. The five grades of stainless steel have been classified according to their microstructure. The first three consist of a single phase but the fourth group contains both ferrite and austenite in the microstructure and the fifth group is precipitation hardenable. As nickel (plus carbon, manganese and nitrogen) promotes austenite and chromium (plus silicon, molybdenum and niobium) encourages ferrite formation, the structure of welds in commercially available stainless steels can be largely based on their chemical composition. Because of the different microstructures, the alloy groups have different welding characteristics and susceptibility to defects. The problems due to welding and their remedial actions are discussed very briefly.

8.2.3.1 Martensitic Stainless Steels

The main problems during welding of martensitic stainless steel are (i) susceptible to cold cracking and (ii) hydrogen pick up due to martensitic structure.

The simple solution is to maintain carbon below 0.25%. High carbon martensitic stainless steels are pre heated to 250°C and cooled slowly in Ms and Mf range to avoid thermal stresses so as to avoid cold cracking. The details regarding slow cooling in Ms and Mf temperatures were discussed in Chapter 2.

To avoid hydrogen pick up use (i) pure argon atmosphere during welding, (ii) low hydrogen electrodes iii) baked electrodes to avoid moisture. Because the final microstructure required is martensite, postweld heat treatment (tempering) is necessary for getting optimum properties. Care should be taken to transform all the austenite to martensite before post weld heat treatment or else austenite will convert to untempered martensite after postweld heat treatment.

8.2.3.2 Austenitic Stainless Steel

As compared to other grades of stainless steels, austenitic stainless steels are easy to weld. Preheat or postweld heat treatment is not necessary and even stress relieving annealing is not required for austenitic stainless steels. However, these grades face major problems of (i) solidification cracking and (ii) sensitization. The solidification cracking is due to high thermal expansion of austenite, that is about 2.5 times higher than ferrite and the austenitic stainless steels are subjected to interdendritic cracking. Introduction of small amount (5–10%) of delta ferrite or free ferrite solves this problem. The weld chemistry is adjusted to have desired delta ferrite as per Schaffer (Figure 8.13) or Delong diagram, that is a plot between austenite and ferrite promoting elements in terms of nickel and chromium equivalents.

In the Delong diagram, the effect of nitrogen is included to calculate the nickel equivalent otherwise it is much similar to the Schaffer diagram. The chemistry of filler metal or the electrode should be adjusted based on the Schaffer diagram to have desired amount of delta ferrite in weld pool to avoid solidification cracking. In short, matching electrodes with base material is not used for welding the austenitic stainless steels. For example, 309 grade electrode is used for welding 304 grade.

The other problem of sensitization is due to segregation of chromium carbides at grain boundaries depleting the chromium content in the near zones as discussed in Section 8.2.2.8. This leads to intergranular corrosion. Sensitization can be avoided by (i) using extra low carbon electrodes, and (ii) by stabilizing the austenite by adding titanium or niobium so that

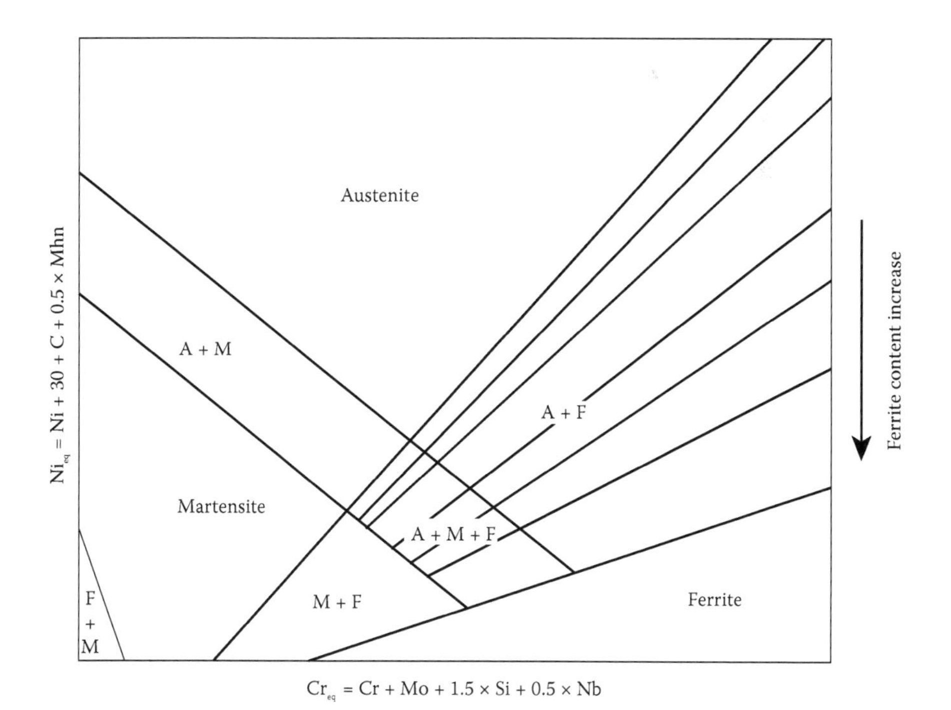

FIGURE 8.13
Schaffer diagram

chromium carbide does not form. The amount of titanium and niobium to be added to avoid the sensitization is calculated from the following empirical equations:

$$Ti = 5 \times (C+N)\ \%$$
$$Nb = 10 \times (C+N)\ \%$$

8.2.3.3 Ferritic Stainless Steel

The major problems in welding of ferritic stainless steels are (i) uncontrollable grain growth (ii) hot tearing and (iii) hydrogen embrittlement. The ferritic stainless steels are "stabilized" by adding titanium and/or niobium to stabilize against grain growth during welding. Hot tearing is due to the presence of impurities such as sulfur, phosphorus, carbon, and its compounds forming unwanted phases mainly during the last stage of solidification. Preheating to 150–200°C helps to reduce the hot tearing by reducing heat input and the holding time at high temperatures thus avoiding precipitation of unwanted phase. The best solution is to eliminate the impurities which is achieved in second and third generation of ferritic stainless steels and these are easily weldable as compared to first-generation ferritic stainless steels. Hydrogen embrittlement can be prevented by using (i) pure argon atmosphere during welding, (ii) low hydrogen electrodes, and (iii) baked electrodes to avoid moisture.

8.2.3.4 Duplex Stainless Steels

Duplex stainless steels have a two phase structure of almost equal proportions of austenite and ferrite. Modern duplex steels are readily weldable but the procedure, especially maintaining the heat input range, must be strictly followed to obtain the correct weld metal structure. Although most welding processes can be used, low heat input welding procedures are usually avoided. Preheat is not normally required and the maximum interpass temperature must be controlled. Choice of filler or electrode is important as it is designed to produce a weld metal structure with a ferrite-austenite balance to match the parent metal. The nitrogen loss during welding is compensated by using the filler or electrode with nitrogen or the shielding gas itself may contain a small amount of nitrogen.

8.2.3.5 Precipitation Hardenable Stainless Steels

Precipitation hardening stainless steels are categorized into three groups: martensitic, semi austenitic, and austenitic. It is important that proper filler metals are used if it is intended that the welds have the same heat treatment response as the base material. Commonly filler materials, having composition close to the welded parts, are used for welding precipitation hardening stainless steels. Weldability of austenitic precipitation hardening stainless steels is poor because of their susceptibility to hot cracks. Limited heat input and welding of parts in solution treated condition are required for diminishing risk of cracks. Nickel alloys (nickel-chromium-iron) are used as filler materials for welding austenitic precipitation hardening stainless steels.

In general, welding of the precipitation hardening stainless steels is similar to the more common non hardenable stainless steels. When proper techniques are used, joints of excellent quality and high strength can be produced in the precipitation hardening alloys. The final heat treatment (postweld heat treatment) is carried out either during or after the joining operation. However, to obtain optimum properties, it is important that the recommended heat treating procedures is followed.

8.2.4 Application for Corrosion Resistance and High Temperatures

Stainless steels are used extensively in industries for their corrosion resistance to both aqueous, gaseous and high temperature environments, for their mechanical properties at all temperatures from cryogenic to the very high, and occasionally for other special physical properties. Uniform corrosion is common in unprotected carbon steels but this does not occur on stainless steels in normal environments and stainless steel will provide unlimited service life without maintenance. Stainless steel is used for buildings for both practical and aesthetic reasons because of its excellent corrosion resistance. Stainless steel was in vogue during the art deco period. Some diners and fast food restaurants use large ornamental panels and stainless fixtures and furniture. Because of the durability of the material, many of these buildings still retain their original appearance. Stainless steel is used today in building construction because of its durability and being a weldable building metal it can be made into aesthetically pleasing shapes. An example of a building in which these properties are exploited is the Art Gallery of Alberta in Edmonton, that is wrapped in stainless steel. Type 316 stainless is used on the exterior of both the Petronas Twin Towers and the Jin Mao Building, two of the world's tallest skyscrapers. The Parliament House of Australia in Canberra has a stainless steel flagpole weighing over 220 metric tons. The aeration building in the Edmonton Composting Facility, the size of 14 hockey rings, is the largest stainless steel building in North America.

Stainless steel is a modern trend for roofing material for airports due to its low glare reflectance to keep pilots from being blinded, also for its properties that allow thermal reflectance in order to keep the surface of the roof close to ambient temperature. The Hamad International Airport in Qatar was built with all stainless steel roofing for these reasons, as well as the Sacramento International Airport in California. Some firearms incorporate stainless steel components as alternatives to blued steel. Some handgun models, such as the Smith & Wesson Model 60 and the

Colt M1911 pistol, are made entirely from stainless steel. This gives a high luster finish, similar in appearance to nickel plating. Unlike plating, the finish is not subject to flaking, peeling, wear-off from rubbing, or rust when scratched.

Martensitic stainless steels have superior wear resistance of high carbon alloys with the excellent corrosion resistance of chromium stainless steels. Martensitic stainless steels have good corrosion resistance to low concentrations of mild organic and mineral acids. Their exposure to chlorides in everyday type activities (e.g., food preparation, sport activities, etc.) is generally satisfactory when proper cleaning is performed after use. These alloys are used where strength, hardness, and/or wear resistance must be combined with corrosion resistance such as cutlery, dental and surgical instruments, nozzles, valve parts, hardened steel balls and seats for oil well pumps, separating screens and strainers, springs, shears, and wear surfaces. Martensitic stainless steels are available as plate, sheet, strip, and flat bars.

The ferritic stainless steels can be made with superior surface finish and with different surface finishes and are most suitable for architectural applications due to its resistance to general corrosion. The refined grades (second and third generation) ferritic stainless steels are used in many fields and applications such as elements and accessories for kitchens and bathrooms, roofs, architectural objects, appliances, safeguards, metallic doors, elevators, storage tanks, and so on. Because of their good thermal conductivity and low thermal expansion these alloys are also used in the manufacturing of chimneys, mufflers, exhaust systems, fasteners, as well as heating elements used in molten salt baths for heat treatments, and so on. Corrosion of unprotected carbon steel occurs even inside reinforced concrete structures as chlorides present in the environment (marine/deicing) diffuse through the concrete. Corrosion products (rust) have a higher volume than the metal and create internal tensions causing the concrete cover to spall. Among the various techniques used to mitigate the corrosion of steel reinforcing bar in concrete, use of ferritic stainless steels or martensitic stainless steels rather than carbon steel is one of the better methods. However, it is not popular due to the high cost.

Stainless steels especially austenitic stainless steels have a long history of application in contact with water due to their excellent corrosion resistance. Applications include plumbing, potable and waste water treatment to desalination. Types 304 and 316 stainless steels are standard materials of construction in contact with water. However, with increasing chloride contents higher alloyed stainless steels such as Type 2205 and super austenitic and super duplex stainless steels are preferred.

Austenitic (300 series) stainless steel, in particular Type 304 and 316, are the materials of choice for the food and beverage industry. Stainless steels do not affect the taste of the product are easily cleaned and sterilized to prevent bacterial contamination of the food, and are durable. However, acidic foods with high salt additions, such as tomato sauce, and highly salted condiments, such as soya sauce may require higher alloyed stainless steels such as 6% molybdenum superaustenitic stainless steel to prevent pitting corrosion by chloride.

Surgical tools and medical equipment are usually made of stainless steel, because of its durability and ability to be sterilized in an autoclave. In addition, surgical implants such as bone reinforcements and replacements (e.g., hip sockets and cranial plates) are made with special alloys formulated to resist corrosion, mechanical wear, and biological reactions in vivo. Stainless steel is used in a variety of applications in dentistry. It is common to use stainless steel in many instruments that need to be sterilized, such as needles, endodontic files in root canal therapy, metal posts in root canal treated teeth, temporary crowns and crowns for deciduous teeth, and arch wires and brackets in orthodontics. The surgical stainless steel alloys (e.g., 316 L) have also been used in some of the early dental implants.

Cookware and bakeware may be cladded in stainless steels for easy cleaning, durability and for use in induction cooking (this requires a magnetic grade of stainless steel, such as 432). Because stainless steel is a poor conductor of heat, it is often used as a thin surface cladding over a core of copper or aluminium to conduct heat more readily. Stainless steel is used for jewelry and watches, with 316L being the type commonly used for such applications. Super austenitic stainless steels with 6% molybdenum are used in bleach plant and Type 316 is used extensively in the paper machine. It can be refinished by any jeweler and will not oxidize or turn black. Valadium, a stainless steel and 12% nickel alloy, is used to make class and military rings. Valadium is usually silver-toned, but it can be electroplated to give it a gold tone. The gold tone variety is known as Sunlite Valadium. Other "Valadium" types of alloys are trade named differently, with names such as "Siladium" and "White Lazon."

Duplex stainless steels are used in heat exchangers, tubes and pipes for production and handling of gas and oil, heat exchangers and pipes in desalination plants, mechanical and structural components for corrosive environments, pipes in process industries handling solutions containing chlorides, utility and industrial systems such as rotors, fans, shafts and press rolls where high corrosion fatigue strength is required – cargo tanks, vessels, piping, and welding consumables for chemical tankers. A wide range of stainless steels are used throughout the paper making process. For example, duplex stainless steels are being used in digesters to convert wood chips into wood pulp.

The stainless steel, however, satisfies the definition for high strength steels in addition to corrosion resistance, they are not widely used because of cost due to the high amount of alloying additions. However, some automotive manufacturers use stainless steel as decorative items in their vehicles. The use of corrugated stainless steel panels popular during the

1960s and 1970s, declined due to high cost and are again picking up in metro trains due to the availability of lean grade austenitic stainless steels such as 200 series, 301, and 304. In addition to good appearance, the lean alloys strain harden rapidly and thus gain strength and wear resistance. The coaches built with these lean grade austenitic stainless steels, on accidents due to impact load, will strain harden and gain strength and will not get crushed ensuring the safety of the passengers. Stainless steels are used in aircrafts during 1930s but the use of stainless steel in mainstream aircraft is hindered by its excessive weight compared to other materials, such as aluminium.

Important considerations to achieve optimum corrosion performance are: choose the correct grade for the chloride content of the water; avoid crevices when possible by good design; follow good fabrication practices, particularly removing weld heat tint and sensitization; avoid formation of strain induced martensite in austenitic stainless steels since even small amount of martensite in austenitic stainless steels (especially in 316 grades) will undergo stress corrosion cracking; drain promptly after hydro testing. Selection of appropriate stainless steel for specific corrosion environment is briefed in Chapter 9.

Hardness and strengths change very little with increases in tempering temperature up to 480°C and so martensitic stainless steels can be used up to temperature range of 400°C. Similarly, the ferritic stainless steels can also be used up to 400°C. The austenitic stainless steels can be used up to a temperature of 600°C. The other advantage in austenitic grade is the absence of ductile to brittle temperature and hence can be used at cryogenic temperature also. The stainless steels are most suitable material where moderate temperature and corrosion coexist.

8.3 Tool Steels

As the name indicates, steels used for making tools for cutting and forming are called "tool steels." The tool should have better properties than the workpiece materials to serve the intended purpose.

8.3.1 Properties Required for Tool Steels

The properties required are:

(i) high room temperature hardness to withstand against wear
(ii) good hardenability for through hardening of thick sections
(iii) resistance to thermal softening to withstand the temperature produced during machining
(iv) good toughness and ductility to withstand shock loads

(v) high stiffness for dimensional tolerances
(vi) good wear resistance for long life
(vii) resistance to chemical reactions with the workpiece and lubricants
(viii) high thermal conductivity to dissipate the heat
(ix) low thermal expansion for maintaining accuracy
(x) good machinability and grindability to shape the tools that are normally complicated
(xi) availability at cheaper cost.

In order to achieve these properties, more numbers of alloying elements in high amounts are added to steels and the necessity and effect of each element is briefly given here.

8.3.2 Effect of Alloying Elements

Carbon is the element that increases both hardness and hardenability and thus increases abrasion (wear) resistance. Carbon is the cheapest alloying element in tool steels. Chromium increases hardenability, wear resistance by forming carbides and corrosion resistance by forming chromium oxide layer. Chromium also increases resistance to molten metal and oxidation useful in making die steels for melting low temperature metals such as aluminium. It also increases scaling resistance. Five percent chromium can resist scaling up to 650°C, 8% chromium is required for scaling resistance up to 750°C. About 10–12% chromium is required to resist temperature of 850°C. Addition of Si (0.8–1%) increases strength of chromium oxide layer and can further increase the resistance to oxidation and scaling. Chromium also protects from chemical attack and liquid metal attack. Vanadium and molybdenum form primary carbides and act as grain refiners apart from increasing hardenability. Silicon increases hardenability and high temperature strength. Nickel increases hardenability and toughness and imparts shock resistance to tool steels and reduces cracking while quenching by decreasing austenitizing temperature. Cobalt improves red hardness (or) hot hardness (resistance to thermal softening) but it decreases hardenability. Normally, reduction in hardenability due to cobalt is compensated by adding other alloying elements that increase hardenability. Tungsten apart from forming primary carbides improves hot hardness.

8.3.2.1 Role of Carbides in Tool Steels

As discussed in Chapter 5, the carbide forming elements form carbides such as M_7C_3, MC, M_6C, or $M_{23}C_6$. The properties, especially the wear resistance in tool steels, are decided by the type of carbides present. The role of carbides is very critical and its impact on hardness and wear resistance is shown in Table 8.8.

The tool steel listed in serial number 3 has low hardness and wear rate due to high volume of carbides and type of carbides. It forms MC type and M_7C_3 type of carbides. The tool steel listed in serial number 7 has the highest hardness but the wear resistance is low although the total carbide content is high. It has low amount of MC type of carbides and forms more M_6C and $M_{23}C_6$ type of carbides. Increasing M_7C_3 from 11–12% to 14–15% the wear resistance can be increased by a factor of two. However, further increase in M_7C_3 causes decrease in malleability. Increasing MC content can further increase the wear resistance but more than 25% decreases grindability. In case of tool steels used for elevated temperatures, wear rate mainly depends on size and distribution of carbides. At high temperature the carbides coalesce and form coarse precipitates decreasing the properties. M_7C_3 carbides easily coalesce than MC type and so for elevated temperatures MC carbides are preferred for better properties. The carbide forming elements should be cautiously selected for the intended application.

8.3.3 Type of Tool Steels

There are many types of tool steels. They are named based on their hardening conditions, properties, and/or applications. A few important types of tool steels are briefed in this section.

They are:

(i) ***Water hardening tool steels*** are represented as "W" followed by numerals (W1–W5). Their carbon content varies between 0.6 and 1.4% with chromium of maximum 0.5% and vanadium 0.25% – simple high carbon steel. These grades are hardened by quenching in water and

TABLE 8.8

Relation between type and volume of carbides with hardness and wear resistance at room temperature

S. No	Hardness (HRc)	Total carbide content (% Vol)	M_7C_3	MC	M_6C and $M_{23}C_6$	Volume of worn off, mm^3
1	61	16–17	16–17	–	–	0.268
2	61.5	17	17	–	–	0.263
3	60	23–25	19–20	4–5	–	0.117
4	60	16–18	9–10	7–8	–	0.228
5	61.5	16–18	9–10	7–8	–	0.223
6	62	22	–	3	19	0.277
7	63.5	22	–	3	19	0.271

hence called water hardening tool steels. They are very hard and strong but lack in ductility and toughness.

(ii) ***Shock resisting tool steels*** are denoted as "S" followed by numerals (S1-S7). The nominal composition in % is: C – 0.45–0.65, Si – 1–2, Cr – 1.5–3.5 and Mo – 0.5–1.4. They have good toughness due to relatively low carbon than water hardening tool steels and can resist shocks or repeated loading and hence known as shock resisting tool steels.

(iii) ***Cold work tool steels (or) Oil hardening steels*** are symbolized as "O" followed by numerals (O1-O7). These grades have manganese (1–1.6%,), silicon (1%), chromium (0.5–0.75%), tungsten (0.5–1.75%), and molybdenum (0.25%), along with carbon (0.9–1.2%). These grades of steels have good hardenability and as martensite forms upon quenching in oil are called oil hardening tool steels.

(iv) ***Air hardening tool steels*** are indicated as "A" followed by numerals (A2 – A4, A6- A10). These grades of tool steels contain high amount of carbon (1–1.35%), manganese (1.8–2%), silicon (1.25%), chromium (1–5%), nickel (1.5–1.8%), vanadium (1–4.75%), tungsten (1–1.25%), and molybdenum (1–1.5%). They can be hardened by cooling in air to very high hardenability due to the alloying elements and hence called air hardening tool steels. These grade of tool steels have excellent stiffness, good wear resistance, and moderate red (or) hot hardness but they have poor fabricability.

(v) ***High Carbon High Chromium steels*** are indicated as "D" followed by numerals (D2 – D5, D7). These grades have high carbon (up to 2.25%), high chromium (12%) along with vanadium (4%), molybdenum (1%), and cobalt (3%). Due to high carbon and chromium they are called high carbon high chromium (HCHCR) tool steels. These grades have excellent wear resistance, good abrasion resistance, and minimal dimensional changes during hardening.

(vi) ***Hot working tool steels*** are specified as "H" followed by numerals. These grade tool steels are used for making tools for high temperature applications such as dies for hot forging, die casting dies, plastic moulding dies, and so on, and hence are called hot working tool steels. Based on their composition, the hot working tools steels are subdivided as chromium base (10–19), tungsten base (21–26) and molybdenum base (41–43). The chrome base hot working tool steels have about 3.25 to 4.25% chromium along with carbon – 0.35–0.4%, vanadium – 0.4–2%, tungsten – 1.5–5% and molybdenum – 1.5–2.5%. The chrome base tool steels have low thermal expansion, good red hardness, good toughness, low distortion while hardening, medium resistance to corrosion and oxidation and good fabrication characteristics. Tungsten base hot working tool steels have tungsten about 9–18% along with 2–12% chromium and 0.25–0.5% carbon. These

grades have low distortion during quenching, good hot hardness, good toughness and fair wear resistance. These grades are costly as compared to chromium base and fabrication properties are inferior to chrome base but can with stand higher temperatures than chrome base. In molybdenum base hot working tool steels, tungsten is partially replaced by molybdenum. The nominal composition (in wt %) of the molybdenum based tool steels are: C – 0.55–0.65, Cr – 4, V – 1–2, W – 1.5–6 and Mo – 5–8. These steels are cheaper than tungsten-based hot working tool steels.

(vii) High speed steels (HSS) are identified as "T" and "M" based on the major alloying element tungsten or molybdenum based respectively. Advent of HSS in around 1905 made a break through at that time in the cutting speed. It can cut four times faster than the carbon steels and are so known as high speed steel. They are complex iron-base alloys of carbon, chromium, vanadium, molybdenum, or tungsten, or combinations thereof, and with substantial amounts of cobalt. The carbon and alloy contents are balanced to give high attainable hardening response, high wear resistance, high resistance to the softening effect of heat, and good toughness for effective use in industrial cutting operations. Commercial practice has developed two groups of cutting materials as tungsten base (T1, T2, T4–T6, T8, T 15) and molybdenum base (M1–M4, M6–M7, M 30, M 33, M 34, M 36, M 43–M 47). The nominal composition (in wt %) of tungsten base HSS are: C: 0.7–1.5, Cr: 4, V: 1–5, W: 12–20, Co: 5–12. Among the various grades T1 is most widely used that has 18% tungsten, 4% chromium and 1% vanadium and more commonly referred to as 18–4–1.

The material shortages and high costs caused by World War II spurred the development of less expensive alloys substituting molybdenum for tungsten. The advances in molybdenum-based high speed steel during this period put them on par with and in certain cases better than tungsten-based high speed steels resulting in the use of M2 steel instead of T1 steel. Molybdenum about 4.5–9.5% replaced the tungsten from 12–20% to 1.5–6.75%. Both tungsten and molybdenum base HSS have good red hardness and wear resistance. Now a days advanced tools like carbide tools, cemented carbide tools, that can cut 4 to 12 times faster than HSS have superseded HSS, HSS flourish still due to its lower cost and good fabrication properties as compared to the advanced cutting tools.

(viii) Mould steels specified as "P" followed by numerals (P1–P39) are used for making casting dies to handle the liquid metal requiring good shock resistance rather than hardness. Hence, the carbon content is kept low (0.07–0.1%). The other alloying elements such as chromium (0.6–5%), nickel (0.5–3.5%), and molybdenum (0.2% max.) are added to increase the sticking resistance and toughness for making moulds.

(ix) ***Special purpose tool steels*** are of two types, low alloy type (L) and carbon-tungsten based (F). The low alloy type has about 0.5–1.1% carbon, 0.75–1.5% chromium, maximum of 1.5% nickel, maximum of 0.2% vanadium and molybdenum of 0.25% maximum. These steels have good wear resistance, toughness and fair resistance to dimensional tolerances. The carbon-tungsten type has tungsten 1.25–3.5% and carbon 1–1.25%. These grades are brittle and have high wear resistance.

8.3.4 Classification Based on Properties

Based on thermal stability, the tool steels are classified as (i) nonthermo stable (W and S type), (ii) semithermo stable (D, O and A types), and (iii) thermo stable (T and M based HSS). Nonthermo stable tool steels achieve high hardness and strength by forming martensite which in turn depends on the amount of carbon. Mostly nonthermo stable tool steels are plain carbon and with low alloying elements. The properties fall if the temperature is increased above 200°C due to tempering of martensite. Semithermo stable tool steels will have high carbon and high carbide forming elements such as chromium, molybdenum, vanadium, tungsten, and so on. The hardness and strength is due to martensite and carbides. The alloy carbides resist the softening of martensite during tempering and can be used for slightly elevated temperatures (about 400°C). For Thermo stable tool steels, the hardness and strength are due to martensite, carbides, and precipitation hardening (or) intermetallics. The carbide formers such as tungsten, molybdenum, and vanadium form also intermetallics along with their respective carbides. Hence, more amounts of these elements are added in the thermo stable tool steels. The strength, hardness, and other properties mainly depend on carbides, their size and distribution. The thermo stable can with stand high temperatures.

8.3.5 Heat Treatment of Tool Steels

Heat treatment of tool steels differs slightly from the general heat treatment discussed in Chapter 2. The tool steels are used in hardened and tempered condition and are stress relieved after rough machining to the required shape. The work sequence is: rough machining, stress relieving, semifinish machining, hardening and tempering, and final grinding to required dimensions.

Stress relieving: This treatment is done after rough machining by heating to 550–650°C. The material is heated until it has achieved uniform temperature all the way through and then cooled slowly, in a furnace.

Heating to hardening (austenitizing) temperature: The fundamental rule for heating should be slow to hardening (austenitizing) temperature to minimize distortion. Normally, vacuum furnaces or furnaces with controlled protective gas atmosphere or molten salt baths are used to avoid decarburization and oxidation that result in low surface hardness with a risk of cracking. Holding time at hardening temperature cannot be generalized to cover all heating situations and the composition of the tool steel. It varies between 30 minutes to few minutes. Especially in HSS, the soaking time should be minimum since the carbides can dissociate and carbon enters austenite and the alloying elements (ferrite stabilizers) go in to solution and try to stabilize ferrite. After quenching it ends up with martensite and retained ferrite that not only reduces the properties and cannot be restored. Quenching rate (critical cooling rate) is decided based on the TTT diagram as discussed in Chapter 2. The choice between a fast and slow quenching rate is usually a compromise; to get the best microstructure and tool performance, the quenching rate should be rapid; to minimize distortion, a slow quenching rate, but higher than critical cooling rate, is recommended. Appropriate quenching medium needs to be selected for correct results and great care has to be taken while quenching through the martensite range, as martensite formation leads to an increase in volume inducing stresses in the material. Martempering is one of the best of all hardening methods.

New concepts have been introduced with modern types of furnaces, and the technique of quenching at a controlled rate in a protective gas atmosphere or in a vacuum furnace with gas is becoming increasingly widespread. The cooling rate is roughly the same as in air for protective gas atmosphere, and the problem of oxidized surfaces is eliminated. Incomplete formation of martensite results in retained austenite on quenching decreasing the properties. Complete formation of martensite should be ensured before tempering treatment. Tool steels that are water- or oil-hardened need subzero cooling to have full martensite because in most cases the Mf temperature is below room temperature. In case of air-hardened tool steels, tempering itself will convert the retained austenite to martensite and this needs double or triple tempering discussed here briefly.

Tempering: The material should be tempered immediately after quenching. If it is not possible, the material must be kept warm, in a special "hot cabinet," awaiting tempering. The choice of tempering temperature is often determined by experience and needs. If maximum hardness is desired, tempering must be done at about 200°C, but never lower than 180°C. High speed steel is normally tempered at about 20°C above the peak of the secondary hardening temperature. If a lower hardness is desired, higher tempering temperature is employed. Normally, double tempering is recommended for tool steel that are air hardened and triple tempering is done for high speed steel with a high carbon content, over 1%.

After quenching (hardened in air), certain amount of austenite remains untransformed when the material is to be tempered. During first tempering, most of the austenite is transformed to martensite and untempered. A second tempering gives the material optimum toughness and required hardness. The same line of reasoning can be applied with regard to retained austenite in high speed steel. In this case, however, the retained austenite is highly alloyed and transformation to martensite is slow. During tempering, some diffusion takes place in the austenite, secondary carbides are precipitated resulting in low alloyed austenite that is more easily transformed to martensite when it cools after tempering. So, third tempering can be beneficial in driving the transformation of the retained austenite further to martensite. Holding times in connection with tempering is also critical. After the tool is heated through, hold the material for at least two hours at full temperature each time.

Apart from the regular heat treatment, surface treatments such as nitriding, nitro carburizing are carried out to increase the surface hardness and wear properties. Surface coating of tool steel is becoming more common. The hard coating normally consists of titanium nitride and/or titanium carbide. These coatings have very high hardness and low friction giving a very wear resistant surface, minimizing the risk of adhesion and sticking. The two most common coating methods are: physical vapour deposition (PVD) coating and chemical vapour deposition (CVD) coatings.

Generally, the tool steels used for cutting operation will have carbon more than 0.6% carbon and are hardened to high hardness (59–60 HRc) values but the tool steels used for making dies will have carbon less than 0.6% and are hardened to low value of hardness (42–50 HRc). The die steels need some toughness to arrest the propagation of cracks formed during operation while the same is not required for cutting grade tool steels as they are under the state of triaxial compressive stresses as against die steels, that experience tensile stresses also.

8.3.6 Applications for Machining and Forming

Water-hardening tool steels with 0.6–0.75% carbon are used where some toughness is required such as hammers and concrete breakers. For water-hardening tool steels with carbon content of 0.75–0.95% used for tools such as punches, chisels, dies, and shear blades hardness is of prime importance rather than toughness. In case of wood working tools, drills, taps, reamers, turning tools, and so on, where wear resistance is more important, water-hardening tool steels with 0.95–1.4% carbon are used. However, the applications are limited to soft materials like wood, brass, aluminium, and soft steels that require low cutting speed and low cost. Shock resisting tool steels are mainly used as forming tools, punches, chisels, pneumatic tools, and shear blades. Cold work tool steels (or) oil-hardening grade tool steels are used for making tools for cold working operations such as taps, form tools, and expansion reamers. These types of tool steels have good wear resistance. Air-hardening tool steels are used for making blanking, forming, trimming, and thread rolling dies. High carbon high chromium steels are used for making blanking, piercing dies, wire drawing dies, bar and tubes, thread rolling dies, and master gauges. Hot working tool steels (chromium-based) are used for making hot extrusion dies, die casting dies, hot forging dies, mandrels, and hot shearing dies. Hot working tool steels (tungsten-based) are used for making mandrels and extrusion dies for brass, nickel alloys, and steels. The applications of hot working molybdenum-based tool steels are same as that of tungsten-based tool steels but are lower cost than tungsten-based tool steels.

High speed steels are used for making cutting tools, tool bits, drills, reamers, broaching taps, milling cutters, hobs, saws, and wood working tools. Mould steels are used for making master hub, low temperature die casting dies, and injection and compression moulding dies for plastics. Special purpose tool steels (low alloy) are used for making bearings, rollers, clutch plates, cam collets and wrenches of the machine tools and also in gauges, knurls, and so on. Special purpose tool steels (C-W) are used for paper cutting knives, wire drawing dies, plug gauges, forming, and finishing tools where toughness is not the major requirement.

9

Selection of Materials

9.0 Introduction

Most problems originate from improper selection of work, material, men, process, tools/equipment, place, and time. If any one of these is wrong the result is undesirable. Material selection is an important step in the process of designing any physical object in the context of product design, safe and reliable functioning of a part or component at minimum cost while meeting product performance goals. Systematic selection of the best material for a given application begins with properties, cost, and other nontechnical parameters of candidate materials. For example, a thermal blanket must have poor thermal conductivity in order to minimize heat transfer for a given temperature difference. The material selected for the blanket should be affordable. The other nontechnical factors such as availability, ordering time, skill, and/or facility available to fabricate the blanket in the local conditions decide the material for making a thermal blanket.

9.1 Tools Used for Selection of Materials

Normal methods of selection of materials are:

(i) consulting experts since they have developed the expertise over long experience and experimentations
(ii) based on the established handbooks
(iii) searching in the Internet.

The main problems in these methods can be:

(i) difficult to get experts, or too costly for small tasks
(ii) details in handbooks may be obsolete in the fast-changing materials world

(iii) very often local conditions may not match with the search results from the Internet or difficult to avoid bogus information.

The other best way is to systematize the process of selection such that even a novice will be able to follow. Objective of systemizing is certainly not to suggest "the best" material for any component: "Name a component – here is the suitable material." But to help arrive at the most appropriate material logically. As the proverb goes, "It is a lot better to teach fishing rather than gifting a pound of fish."

9.2 Systemized Selection of Materials

The rudiments of selection are the same whether the selection is engineering material or men or process – or, for that matter, anything. If the basics of selection are understood, the best possible material, process, or men can be selected for every work. In order to systematize, three groups of data are required: (i) data on availability, (ii) data on requirements, and (iii) data on nonfunctional aspects.

The data on availability includes the different engineering materials available in that location, their technologically relevant properties, manufacturing properties and supply conditions such as minimum quantity, forms of availability (sheets or rods or powders, and so on.), delivery schedules, and so on. Engineering materials can be broadly classified as metals such as iron, copper, aluminum, and their alloys, and so on, and nonmetals such as ceramics (e.g., alumina and silica carbide), polymers (e.g. poly vinyl chloride), natural materials (e.g., wood, cotton, flax, and so on.), composites (e.g., carbon fibre reinforced polymer, glass fibre reinforced polymer, and so on.), and foams. Each of these materials is characterized by a unique set of physical, mechanical, and chemical properties, that can be treated as attributes of a specific material. The selection of material is primarily dictated by the specific set of attributes required for an intended service. In particular, the selection of a specific engineering material for a part or component is guided by the function it should perform and the constraints imposed by the properties of the material. For example, electrical wires need good electrical conductivity to avoid electrical losses. Since metals are good conductors of electricity metals can be considered for selection. Among the metals silver, copper, aluminium, steel, or nickel can be considered. The data on manufacturing properties needed to shape the selected material for the desired shape or contour as well as the manufacturing properties such as machinability, formability, weldability, and castability will be required.

Next data required will be functional requirements such as strength, hardness, wear resistance, and so on, service life expected, shaping required, cost of service failure and service environment such as temperature, corrosion, humidity, and so on. The technical requirements are guided by the design. The service life expected is more crucial in selection as it varies from few seconds to few years. The material selection is mainly based on the service life. For example, the passenger car's service life is about fifteen years. The material for body in white should be selected accordingly. High corrosion resistant materials like stainless steels and titanium although suitable, these metals will increase the cost of the automobile. A coated/painted steel can comfortably withstand fifteen years and, hence, it is better to select any of the suitable grade of steel based on other requirements. The shape and size of the component is decided by the design and the selected material should have the manufacturing properties. Cost of service failure plays a major role in the selection of material and the cost of service failure is taken care by the factor of safety in design. For example, failure of the aerospace components during service will result in heavy damages. With a view to avoid service failure the factor of safety can be increased, that will increase the initial cost and running (fuel) cost. The best alternative is to use the material with stringent norms in the required properties, normally called as aerospace grades, will reduce the probability of failure within the factor of safety. Service environment is another important parameter to consider. For example, in urban conditions, weathering steels is best against corrosion resistance but it cannot be used in industrial or marine conditions due to the presence of high amount of sulfur and chlorine in the industrial and marine environment.

The third data required are on aspects such as cost, legality, conventional/traditional preferences, and aesthetics. Cost is more important especially in comparison with the market competitors. The purchasing power of the people has great impact on the cost. Legality and environmental issues should also be considered which vary from place to place and country to country. Conventional/traditional preferences and aesthetic tastes of the clients are more important in the consumer products. For example, colored stainless steel though technically success, the products made of colored stainless steel have not succeeded commercially mainly because of tradition that stainless steels must have a shiny white surface.

A judicious combination of all these three factors leads us to the selection of appropriate materials. Very often more than one material may meet the requirements. In such cases, selection may be based on the ranking of the functional and nonfunctional requirements.

9.3 Case Studies

9.3.1 Case I: Manufacturing of Gears

Data on availability: Assume practically all types of materials are available in the region and also their technical properties, supply conditions, and so on, are known. So the first data group is ready.

Data on requirements: The gears are for torque transmission. The magnitude of torque transmitted should be known. If it is minimal as in clocks, even polymeric (plastic) gears would be tolerable. For slightly higher level as in car windshield wiper motor, medium carbon steel will be suitable. For the torque for driving a car, low alloy steel in heat treated condition may be required and so on and so forth. The service life expected depends on the product, say for clocks, five years is sufficient whereas for automobiles it should be about fifteen years. The dimensions and shape of the gear, decided by the design engineer, depends on facility available in local conditions. Cost of service failure is high for automobiles but for clock it is low. Service environment like local temperature conditions, corrosive environments should also be considered. While selecting the gear material for winter conditions and marine atmospheres, the material behavior with respect to temperature and its corrosion resistance need to be prioritized.

Data on nonfunctional aspects: Cost is the major influencing factor in this group. Based on the cost (raw material and manufacturing) plastics, mild steel, gun metal, and alloy steels are suitable in the increasing order. Based on the cost of failure and service environments, the material can be selected. If these factors are low, cheaper material such as plastics can be selected. No other nontechnical factor is serious in this example.

9.3.2 Case II: Selection of Reinforcing Material for Cement Concrete in Marine Environments

Data on availability: Assume practically all types of materials are available in the region and also their technical properties, supply conditions, and so on, are known. So the first data group is ready.

Data on requirements: The reinforcement should withstand the static load, the thermal expansion should match with concrete and it should resist corrosion due to chlorides; the service life and maintenance of the concrete.

Data on nonfunctional aspects: Cost is the major influencing factor in the selection. No other nontechnical factor is serious in this example.

Selected materials: Based on the static strength, thermal expansion and cost, most suitable material is mild steel. But the corrosion of unprotected carbon steel occurs even inside reinforced concrete structures as chlorides present in the environment (marine) diffuse through the concrete. Corrosion products (rust) having a higher volume than the metal create internal tension causing the concrete cover to spall. Mitigating the corrosion of steel reinforcing bar in concrete is a must. Various techniques are recommended: thicker concrete cover; cathodic protection; membranes, epoxy coatings, and so on.

The other alternative is to use stainless steel rather than carbon steels. Stainless steel provides both strength and corrosion resistance inside the concrete, providing a long, maintenance free service life of the structure but the cost is higher as compared to protected mild steel. It should be noted that austenitic stainless steel should not be selected because its thermal expansion being two and half time higher than concrete will damage the concrete from inside due to temperature fluctuations.

Among these two materials, the selection is based on which factor is prioritized: cost or the maintenance-free service life. If it is cost, coated mild steel is suitable material; if it is maintenance-free service life, stainless steel is the appropriate material.

9.3.3 Case III: Selection of Stainless Steel for Architectural Applications

In general, all grades of stainless steel have good resistance to general corrosion compared to carbon steels. But for specific environments, correct grade of stainless steel has to be selected for maintenance-free long service life with a low life cycle cost and excellent sustainability. It is important to note that this selection is only for appearance and not for structural integrity.

Data on availability: Assume practically all types of stainless steels are available in the region and also their technical properties, supply conditions, and so on are known. So the first data group is ready.

Data on requirements: The main requirement is maintenance free, sustainable grade of stainless steel at low cost for the given environment. So, the following factors have to be considered: (i) environmental pollution, (ii) coastal exposure or deicing salts exposure, (iii) local weather pattern, (iv) design considerations, and (v) maintenance schedule.

The criteria required for this application should be evaluated based on the score for each of the conditions as given in Table 9.1.

Selected materials: Based on the scores, the grade of stainless steels is selected. If the score is between 0 and 2, 304 or 304 L will be an economical choice. Type 316/316L or 444 is the most economical choice if the score is 3. Type 317L or a more corrosion resistant stainless steel is suggested for the score 4. If the score is 5 and above, more corrosion resistant stainless steel such as 4462, 317LMN, 904L, superduplex, superferritic, or a 6% molybdenum superaustenitic stainless steel is to be selected.

Systematic selection for applications requiring multiple criteria is more complex. For example, a rod that should be stiff and light requires a material with high Young's modulus and low density. If the rod will be pulled in tension, the specific modulus, or modulus divided by density, will determine the best material. But because a plate's bending stiffness scales as its thickness cubed, the best material for a stiff and light plate is determined by the cube root of stiffness divided by density. An Ashby plot, named for Michael Ashby of Cambridge University, is a scatter plot that displays two or more properties of many materials or classes of materials. These plots are useful to compare the ratio between different properties. For the example of the stiff/light part discussed above an Ashby plot will have Young's modulus on one axis and density on the

TABLE 9.1

Points for each functional requirements

Points	Criteria – Environmental pollution
0	Very low or no pollution
1	Low (urban conditions)
2	Moderate (urban conditions)
3	High (urban conditions)
3	Low or moderate (industrial conditions)
4	High (industrial conditions)
Criteria – Coastal exposure	
1	Low (1.6 to 16 km from salt water)
3	Moderate (30 m to 1.6 km from salt water)
4	High (< 30 m from salt water)
5	Marine (some salt spray or occasional splashing)
8	Severe marine (continuous splashing)
10	Severe marine (continuous immersion)
Deicing salts exposure	
0	No salt was detected on a sample from the site and no change in exposure conditions is expected or traffic and wind levels on nearby roads are too low to carry chlorides to the site and no deicing salt is used on sidewalks
1	Very low salt exposure (≥ 10 m to 1 km)
2	Low salt exposure(< 10 to 500 m)
3	Moderate salt exposure (< 3 to 100 m)
4	High salt exposure (< 2 to 50 m)
Local weather pattern	
-1	Cold climates, regular heavy rain, humidity below 50%
0	Tropical or subtropical, wet, regular or seasonal very heavy rain
1	Regular very light rain or frequent fog, humidity above 50%
2	Hot, humidity above 50%, very low or no rainfall
Design Considerations	
0	Boldly exposed for easy rain cleaning and vertical surfaces
-2	Surface finish is pickled, electro polished, or roughness ≤ Ra0.3 μm
-1	Surface finish roughness Ra 0.3 μm < X ≤ 0.5 μm
1	Surface finish roughness Ra 0.5 μm < X ≤ 1 μm
2	Surface finish roughness Ra > 1 μm
1	Sheltered location or unsealed crevices
1	Horizontal surfaces
Maintenance schedule	
0	Not washed
-1	Washed at least naturally
-2	Washed four or more times per year
-3	Washed at least monthly

other axis, with one data point on the graph for each candidate material. On such a plot, it is easy to find out not only the material with the highest stiffness, or that with the lowest density but also with the best ratio. Using a log scale on both axes facilitates selection of the material with the best plate stiffness.

Subjective Questions

Short Answers

1. Why is steel known as a moving standard?
2. What are the four allotropes of iron?
3. Steel is an enigma – it rusts easily, yet it is the most important of all metals. Justify this statement.
4. Why are phase transformations involving austenite very important in the heat treatment of steels?
5. Why is the iron-carbon diagram not a true equilibrium diagram?
6. Although the iron-carbon diagram is not a true equilibrium diagram, it is useful for practical applications. Why?
7. Write down the peritictic reaction in the iron-carbon diagram. Mention the concentration of carbon and temperature at which it occurs.
8. Write down the eutectic reaction in the iron-carbon diagram. Mention the concentration of carbon and temperature at which it occurs.
9. Write down the eutectoid reaction in the iron-carbon diagram. Mention the concentration of carbon and temperature at which it occurs.
10. What is pearlite?
11. Define heat treatment.
12. What are the four basic types of heat treatment?
13. Why is the strength of the normalized steel higher than annealed steel of the same composition?
14. Define hardening.
15. Why won't the phase formed due to hardening be found in the iron carbon diagram?
16. What is meant by transformation diagrams? What are the two types of transformation diagrams?
17. Define critical cooling rate.
18. What are the two types of martensite? State the relation between the amount of carbon and the type of martensite formed.
19. Schematically represent the relationship between carbon content and the maximum obtainable hardness in steels.

20. Define hardenability.
21. List the quenchants used for hardening the steels.
22. What is the purpose of tempering the hardened steels?
23. What are the types of isothermal heat treatments?
24. What is meant by surface hardening?
25. What are the two types of surface hardening?
26. What is the purpose of adding alloying elements in steel?
27. List the ferrite and austenite stabilizers in steels.
28. Which are the alloying elements that stabilize ferrite without forming carbides?
29. Why cobalt is called as a neutral stabilizer?
30. Phosphorous is the most effective solid solution hardener but it will be normally restricted below 0.25% in steels. Why?
31. What are the elements present in steel as metallic particles?
32. What is meant by subzero cooling? Why it is necessary?
33. Define carbon equivalent.
34. What is the amount of carbon in 1060 steel?
35. Nickel is the first element alloyed with iron, but has been removed now from the series of low alloy steels. Why?
36. Why has the use of tungsten in low alloy steels declined since 1940?
37. Why are silicon steels called electric steels?
38. Why does boron facilitate isothermal processes in steels?
39. Define high strength steels.
40. What are the different types of conventional high strength steels?
41. What are the different types of advanced high strength steels?
42. What is Giga Pascal steel?
43. Why aren't austenitic stainless steel and maraging steels listed as high strength steel though they have high strength and ductility?
44. What is the need for high strength steels?
45. Why aren't conventional high strength steels recommended for electroplating?
46. What is meant by patenting of steel wires?
47. Why do thermomechanical processes have limitations in industrial scale production?
48. The major setback in the nanograined material is the lack of ductility due to the loss of work hardening capacity. How it can be solved?
49. What are the potential advantages of the mixed microstructure of bainitic ferrite and austenite?
50. Why aren't high strength nanobainitic steels as popular as quenched and tempered martensitic steels?
51. Why are maraging steels classified as ultra-high strength steel?
52. Which element is added in maraging steels to induce precipitation hardening?

53. What is meant by marforming?
54. Why aren't maraging steels used in automotive industries?
55. How do stainless steels achieve their "stainless" properties?
56. More chromium is added in stainless steels than required amount for forming self-healing chromium oxide layer. Why?
57. How are stainless steels classified and what are they?
58. Ferritic stainless steels are used for their anticorrosion properties rather than for their mechanical properties: true or false? Justify your answer.
59. What are the types of embrittlement associated with ferritic stainless?
60. Which type of stainless steel(s) can be used for cryogenic applications? Why?
61. What are the classifications of austenitic stainless steels?
62. The formation of martensite is thermodynamically possible in austenitic stainless steels, but it is not forming. Why?
63. Why are austenitic stainless steels quasielastic? What are its implications?
64. What is meant by sensitization in austenitic stainless steels?
65. Why chromium carbides normally precipitate along grain boundaries in austenitic stainless steels?
66. What is meant by stabilization in austenitic stainless steels?
67. In general, what are the weaknesses of austenitic stainless steels?
68. How duplex stainless steels have both ferrite and austenite at room temperature?
69. What are the benefits of duplex stainless steels?
70. What are the three types of precipitation hardenable stainless steels?
71. What are the alloying elements added in precipitation hardenable stainless steels to achieve precipitation hardening characteristics?
72. What are the main problems during welding of martensitic stainless steel?
73. What are the main problems during welding of austenitic stainless steel?
74. What are the main problems during welding of ferritic stainless steel?
75. Why electrodes matching with the base material are not used for welding austenitic stainless steels?
76. What are the advantage and limitation in using ferritic stainless steels or martensitic stainless steels instead of carbon steel in reinforced concrete?
77. Why aren't austenitic stainless steels used in reinforced concrete?
78. Although stainless steels satisfy the definition for high strength steels in addition to corrosion resistance, they are not classified under high strength steels. Why not?
79. List the properties required for tool steels.
80. Although currently, advanced tools such as carbide tools and cemented carbide tools have superseded HSS, still HSS is used. Why?
81. Classify tool steels based on their thermal properties.
82. In heat treatment of high speed steels, soaking time at the austenitic region should be minimum. Why?

83. Generally, tool steels used for cutting operations are hardened to high hardness (59-60 HRc) values while tool steels used for making dies are hardened to low value of hardness (42-50 HRc). Why?
84. What are the normal methods of selection of materials?
85. What are the rudiments of selection of materials?
86. For most of application steel is the preferred material. Why?
87. Define "Cost of service failure"?

Long Answers

1. Draw the iron-carbon diagram, label the phases and lines, and explain the three invariant reactions of the system.
2. Explain the solubility of carbon and properties of ferrite, austenite and cementite phase in iron carbon diagram.
3. Explain the microstructure developed during slow cooling of eutectoid steel.
4. Explain the microstructure developed during slow cooling of hypo eutectoid steel.
5. Explain the microstructure developed during slow cooling of hyper eutectoid steel.
6. What is annealing? Explain various types of annealing with respect to steel?
7. What is normalizing? Explain why the amount of phases differ from the equilibrium diagram when normalized?
8. Represent various types of annealing and normalizing for hypo and hyper eutectoid steels schematically and explain.
9. Explain hardening with respect to TTT diagram.
10. With a schematic sketch, explain how hardenability is measured using the Jominy end quench test?
11. Explain with schematic sketch (i) martempering, (ii) ausforming, and (iii) austempering.
12. Explain the mechanism of precipitation hardening.
13. Explain the various case hardening techniques.
14. Explain case carburizing? Narrate its advantages over other chemical surface hardening methods.
15. Explain case nitriding? Detail its advantages over other chemical surface hardening methods.
16. Explain carbonitriding? Detail its advantages over other chemical surface hardening methods.
17. Explain boriding? What are its advantages over other chemical surface hardening methods?
18. Explain the effect of carbon on mechanical properties of steel.

19. What are the applications, properties, and microstructure of low carbon steels.
20. Explain the applications, properties, and microstructure of medium carbon steels.
21. What are the applications, properties, and microstructure of high carbon steels.
22. Enumerate the effects of the alloying elements on iron carbon diagram.
23. What are the effects of manganese, nickel, and nitrogen on steels?
24. It is necessary to carefully control, not only the nitrogen content, but also the form in which it exists, in order to optimize impact properties. Explain the statement.
25. Discuss the effect of cobalt addition in steels.
26. Discuss the type of carbides formed and their properties in steels when chromium, molybdenum, tungsten, tantalum, vanadium, niobium, zirconium, and titanium added as alloying elements.
27. Summarize the effect of various types of carbides on properties of low alloy steels.
28. What are the elements present in steel as metallic particles? Explain the purpose of adding those elements.
29. Explain the role of manganese, tantalum, titanium, and zirconium in controlling the sulphide and oxysulphides inclusions and its influence on the properties of steel.
30. Explain the effect of alloying elements on hardenability.
31. Explain why a time delay in subzero treatment does not result in complete transformation of retained austenite to martensite?
32. Explain the factor affecting hardenability.
33. Explain the effect of alloying elements on tempering.
34. Write short notes on any two types of low alloy steels.
35. Narrate the microstructure and applications of interstitial free steels.
36. Schematically explain the high strength and advanced high strength steels.
37. Write briefly about any two types of high strength steels and problems in developing those steels.
38. How standard low alloy steel (AISI 4340) is modified as high strength steels? What are its limitations?
39. Write briefly about rephosphorized steel.
40. Write short notes on micro alloyed or high-strength low alloy (HSLA) steels.
41. Enumerate about bake hardening (BH) steels.
42. Explain how patented steel wires achieve their high strength?
43. Write briefly about nanostructured steels.
44. Write short notes on bainitic steels.
45. Explain the conditions required to achieve strong and tough bainitic steels.

46. Explain the metallurgy of maraging steels.
47. Explain the effect of composition on maraging steels.
48. The increase in strength of maraging steels is much higher if both cobalt and molybdenum are added together rather than individually. Why?
49. Write briefly about types of maraging steels.
50. Write short notes on the mechanical properties of maraging steels.
51. Explain the effect of solutionizing and aging temperature on mechanical properties of maraging steels.
52. Narrate about the corrosion behavior of maraging steels.
53. Write briefly the advantages and applications of maraging steels.
54. Write briefly on the effects of alloying elements on structure and properties of stainless steels.
55. How stainless steels are classified? Explain microstructure, properties and applications of any one type of stainless steels.
56. Explain microstructure, properties, and applications of any one type of heat treatable stainless steels.
57. Write briefly about three generations of ferritic stainless steels.
58. Explain the types of embrittlement associated with ferritic stainless.
59. Explain the different types of austenitic stainless steels?
60. Write briefly about strain induced martensite in stainless steels.
61. Discuss the precipitation of carbides in austenitic stainless steels and its effect on properties.
62. What is sensitization? Explain how it can be avoided in austenitic stainless steels.
63. Explain the heat treatment (precipitation hardening) sequence for semi-austenitic stainless steels.
64. Discuss briefly about weldability of stainless steels.
65. Write an essay about application of stainless steels with respect to its corrosion resistance and high temperatures.
66. Discuss the effect of alloying elements on tool steels.
67. Explain the role of carbides in tool steels.
68. Write briefly on the different types of tool steels.
69. Write briefly the complexities of the heat treatment of tool steels.
70. Why is double tempering recommended for tool steels that are air hardened, while triple tempering is done for high speed steel with a high carbon content, for example, over 1% ? Explain briefly.
71. Discuss the applications of tool steels.
72. What are the normal methods of selection of materials? Explain their relative limitations in brief.
73. What are the rudiments of selection of materials and explain them briefly.
74. What are the implications of a cost of service failure in the selection of materials?
75. Explain with an example the role of nonfunctional (nontechnical) aspects in the selection of materials.

Objective Questions

1. Solubility of carbon in FCC-gamma iron is more than in alpha-BCC iron, though the packing factor of FCC is more than BCC
   (a) lower free energy (b) larger void size
   (c) smaller atomic radii (d) more vacancy

2. Which of the following get diffused in the surface layer of the steel parts subjected to nitriding
   (a) Monoatomic nitrogen (b) molecular nitrogen
   (c) Ammonia (d) none of these

3. The maximum solubility of Carbon (in wt %) in FCC iron at 1147°C is
   (a) 0.8 % (b) 2.1%
   (c) 0.025% (d) 6.67%

4. Eutectoid temperature in iron carbon diagram occurs at
   (a) 723°C (b) 1333°F (c) 996 K (d) all

5. From the following microstructures (figure 2.2) of pearlite, identify normalized and annealed (both the steels are having the same composition and at the same magnification)
   (a) I is annealed and II is normalized (b) I is normalized and II is annealed
   (c) both are normalized (d) both are annealed

6. The order of decreasing weldability among the following steels is
   P) Fe- 0.6 C Q) Fe-0.4 C R) HSLA
   a) R-Q-P b) P-Q-R c) Q-P-R d) Q-R-P

7. In steels, which of the following element does not form carbide and stabilizes ferrite
   (a) Ni (b) Mn (c) Si (d) Hf

8. Sample A is austenitized at 900°C and sample B is austenitized at 1300°C. Both the samples are having the same composition and quenched in the same medium. Which will have more depth of hardness?
   (a) Sample A (b) sample B
   (c) both will have same depth of hardness (d) cannot say

9. Of the following which element will increase the austenitising temperature without forming carbides in steels?
   (a) Cr (b) Si (c) Ni (d) Mn

10. During heat treatment:
    (a) Microstructure of a given material is modified, and consequently engineering properties are modified.
    (b) There is no change in microstructure, but properties are modified.
    (c) There are changes only in dimensions.
    (d) None of the above

11. Pearlite is a
    (I) single phase (II) mixture of two phases (III) eutectoid mixture
    (a) I is correct (b) I and II are correct
    (c) II and III are correct (d) II is correct
12. The strength of normalized steel is______________ annealed steel of the same composition
    (a) lower than (b) same as (c) higher (d) cannot say
13. The purpose of tempering
    (a) to convert retained austenite to martensite (b) to soften the martensite
    (c) to harden the martensite (d) to convert austenite to ferrite.
14. Cobalt is a
    (a) austenite stabilizer (b) ferrite stabilizer
    (c) neutral stabilizer (d) carbide former
15. In steels, copper is present as a
    (a) carbide (b) solid solution in ferrite
    (c) solid solution in austenite (d) metallic element
16. The average amount of carbon in 1060 steel is
    (a) 0.6 at% (b) 0.6 wt% (c) 1.06 at% (d) 1.06 at%
17. Of the following, which is not a high strength steel?
    (a) TRIP steels (b) dual phase steels
    (c) stainless steels (d) martensitic steels
18. The element added in maraging steels to induce precipitation hardening is
    (a) manganese (b) molybdenum (c) cobalt (d) titanium
19. The element added in stainless steels to achieve its "stainless" property is
    (a) nickel (b) chromium (c) molybdenum (d) titanium
20. Of the following, which type of type of stainless steel can be used for cryogenic applications?
    (a) ferritic (b) martensitic (c) duplex (d) austenitic
21. In the heat treatment of high speed steels, soaking time at austenitic region should be
    (a) minimum (b) maximum
22. The amount of pearlite in 0.6% carbon steel is
    (a) 75% (b) 9% (c) 25% (d) 91%
23. The amount of ferrite in 0.2% carbon steels is 75%. If the steel is normalized, the amount of pearlite will be (approx.)
    (a) 25% (b) 75% (c) 35% (d) 65%

24. Annealing temperature of hypereutectoid steels is 50°C above
    (a) A1 line (b) A2 line (c) A3 line (d) Acm line
25. Final structure of austempered steel
    (a) pearlite (b) ferrite + graphite (c) bainite (d) martensite
26. Which of the following element makes steel strong at cheaper cost and effectively?
    (a) chromium (b) carbon (c) calcium (d) chlorine
27. Annealing improves
    (a) grain size (b) ductility (c) electrical properties (d) All of above
28. A peritectic reaction is defined as
    (a) two solids reacting to form a liquid (b) two solids reacting not to form a liquid
    (c) liquid and solid reacting to form another solid (d) two solids reacting to form a third solid
29. Austempering is the heat treatment process used to obtain higher
    (a) hardness (b) toughness (c) brittleness (d) ductility
30. The hardness obtained by hardening process does not depend upon
    (a) carbon content (b) work size
    (c) atmospheric temperature (d) quenching rate
31. If the alloy steel at room temperature is magnetic, which phase should not be present?
    (a) ferrite (b) pearlite (c) austenite (d) cementite
32. Which of the following affects the hardenability of steel?
    (a) amount of carbon in austenite (b) austenitic grain size
    (c) insoluble particles in austenite (d) all of the above
33. The minimum carbon percentage required in steel so as to respond to hardening by heat treatment is
    (a) 0.02% (b) 0.08% (c) 0.2% (d) 0.8%
34. The crystal structures of two alloy steels having nominal carbon content
    (i) 18%Cr (ii) 18%Cr8%Ni.
    (a) FCC and BCC (b) both BCC (c) both FCC (d) BCC and FCC
35. Which alloy steel having nominal carbon content (i) 18%Cr and (ii) 18% Cr8%Ni can be hardened by conventional quenching from high temperatures?
    (a) i only (b) ii only
    (c) i and ii (d) neither can be hardened by quenching
36. The hardness of martensite in normal steel is primarily due to
    (a) carbon (b) austenite stabilizers
    (c) ferrite stabilizers (d) neutral stabilizers

37. The crystal structure of martensite in maraging steels is
    (a) BCT (b) BCC (c) FCC (d) HCP
38. Melting point of plain carbon steel is normally
    (a) less than melting point of pure iron
    (b) more than melting point of pure iron
    (c) same as pure iron
    (d) depends on carbon content
39. Which one of the following techniques does NOT require quenching to obtain final case hardness?
    (a) flame hardening (b) induction hardening
    (c) nitriding (d) carburizing
40. A 0.4% plain carbon steel sheet is heated and equilibrated in the inter-critical region followed by instant water quenching. The microstructure of the quenched steel sheet consists of
    (a) fully martensite (b) proeutectoid ferrite + martensite
    (c) martensite + pearlite (d) martensite + austenite
41. The intergranular corrosion can be prevented using
    (a) stabilized grade of stainless steel containing titanium and niobium as alloying elements
    (b) low carbon grade of stainless steel
    (c) both a. and b.
    (d) none of the above
42. Which element precipitates at the grain boundaries, when austenitic stainless steel is heated at 900°C?
    (a) aluminium carbide (b) chromium carbide
    (c) magnesium carbide (d) molybdenum carbide
43. Which of the following is the last phase obtained after completing heat treatment cycle in patenting process?
    (a) bainite (b) martensite (c) pearlite (d) none of the above
44. Which of the following statements is false for heat treatment processes?
    (a) martempering process is designed to overcome limitations of quenching
    (b) pearlite is obtained as the final phase in martempering process
    (c) water is used as quenching medium in the Jominy end quench test
    (d) martensite in maraging steels can be cold rolled
45. Which of the following factors increase hardenability of a steel?
    (a) alloying elements (b) fine grain size
    (c) high nitrogen content in steel (d) all of the above
46. In which of the following methods, surface of a steel component becomes hard due to phase transformation of austenite to martensite?
    (a) nitriding (b) flame hardening (c) both a and b (d) none of the above
47. The process of decomposing martensitic structure, by heating martensitic steel below its critical temperature is called as
    (a) austenitising (b) quenching (c) tempering (d) subzero cooling

48. What is the crystal structure of δ-ferrite?
    (a) body-centered cubic structure (b) face-centered cubic structure
    (c) orthorhomic crystal structure (d) face-centered tetragonal structure

49. Austenite phase in Iron-Carbon equilibrium diagram
    (a) is face-centered cubic structure (b) is magnetic phase
    (c) exists below 727°C (d) is hardest phase

50. The best and economical method for selection of material is
    (a) consulting experts since experts have developed the expertise over long experience and experimentations
    (b) based on established handbooks
    (c) searching the Internet
    (d) to systematize the process of selection

Q.No	1	2	3	4	5	6	7	8	9	10
Answer	b	a	b	d	a	a	c	b	b	a
Q.No	11	12	13	14	15	16	17	18	19	20
Answer	b	c	b	c	d	b	c	d	b	d
Q.No	21	22	23	24	25	26	27	28	29	30
Answer	a	a	c	a	c	b	d	c	b	c
Q.No	31	32	33	34	35	36	37	38	39	40
Answer	c	d	c	d	d	a	b	a	c	b
Q.No	41	42	43	44	45	46	47	48	49	50
Answer	c	b	c	b	a	a	c	a	a	d

Bibliography

1. Venkatsky, S. 1981. *Tales about Metals*. MIR Publishers, Moscow.
2. Budinski, K. G. 2000. *Engineering Materials – Properties and Selection*. Prentice Hall of India Private Limited, New Delhi.
3. Clark, D. S. and W. R. Varney. 1962. *Physical Metallurgy for Engineers*. Affiliated East West Press Private Limited, New Delhi.
4. Avner, S. H. 1974. *Introduction to Physical Metallurgy*. McGraw-Hill Book Company, Singapore.
5. Angelo, P. C. and B. Ravisankar. 2019. *Periodic Table of Elements*. Mahi Publishers, Ahmedabad, 2019, ISBN: 978-81-940137-1-6.
6. Bain, E. C. and H. W. Paxton. 1966. *Alloying Elements in Steel*. ADM, Metals Park, OH.
7. Calister, W. D. 2001. *Fundamentals of Materials Science and Engineering*. John Wiley & Sons, New York.
8. Smallman, R. E. and R. J. Bishop. 1999. *Modern Physical Metallurgy and Materials Engineering*. Butterworth-Heinemann, Oxford.
9. Abbaschian, R., L. Abbaschian and R. E. Reed-Hill. 2000. *Physical Metallurgy Principles*. Cengage Learning, Stamford, CT.
10. Prabha, B., P. Sundaramoorthy, S. Suresh, S. Manimozhi and B. Ravisankar. 2009. Studies on stress corrosion cracking of super 304H Austenitic stainless steel. *Journal of Materials Engineering Performance* (December). DOI: 10.1007/s11665-008-9347-9. https://link.springer.com/article/10.1007/s11665-008-9347-9.
11. Ajith, P. M., P. Sathiya, K. Gudimetla and B. Ravisankar. 2013. Mechanical, metallurgical characteristics and corrosion properties of equal channel angular pressing of duplex stainless steel. *Advanced Materials Research* 717: 9–14.
12. Mahata, A. K., U. Borah, A. Davinci, S. K. Albert and B. Ravisankar. 2014. Room temperature torsional behavior of 15-Cr-15Ni titanium modified austenitic stainless steel. *Procedia Engineering* 86: 166–172.
13. Kondaveeti, C. S., S. P. Sunkavalli, D. Undi, L. V. Hanuma Kumar, K. Gudimetla and B. Ravisankar. 2017. Metallurgical and mechanical properties of mild steel processed by Equal Channel Angular Pressing (ECAP). *Transactions of the Indian Institute of Metals* 70: 83–87.
14. Ravisankar, B. and A. Rajadurai. 1992. Pack boronising of AISI 304 & 410 stainless steel. International Convention on Surface Engineering, INCOSURF 92, Indian Institute of ScienceBangalore.
15. Arivazhagan, R., B. Prabha, S. Manimozhi, S. Suresh, G. Umashanker, R. Nagalakshmi and B. Ravisankar. 2007. Metallurgical studies in super 304H

Austenitic stainless steel. International Symposium on Advances in Stainless Steels: 114–118.

16. Ramakrishnan, S. S., P. Gopalakrishnan, N. Rajeev Kumar, C. S. Venkateswaran and B. Ravisankar. 1996. Molten salt electrolytic boriding of mild steel. Paper Presented at 34th National Metallurgists' Day, New Delhi, India.
17. Sreenivas, T., N. Sakthivel, B. Ravisankar and S. S. Ramakrishnan. 1996. A database management system for selection of steel. Proceedings of the Conference on Computer Applications in Metallurgy and Material Science – CAMME 96: 210–221.
18. Ravisankar, B. and M. Sundar. 2002. Phase transformations in deep drawing of austenitic stainless steel drum – A case study. Proceedings of the National Conference on Processing of Metals: 217–222.
19. Nagalakshmi, R., S. Manimozhi, S. Suresh, B. Prabha and B. Ravisankar. 2007. Corrosion aspects of super 304 H material and its weldments. National Conference on Corrosion and Its Control: 1–5.
20. Bhat, S. P. Advances in high strength steels for automotive applications. Automotive Product Applications, ArcelorMittal Global R&D, East Chicago. www.autosteel.org/-/media/files/autosteel/great-designs-in-steel/gdis-2008/12—advances-in-ahss-for-automotive-applications.ashx (accessed April 21, 2018)
21. Tamarelli, C. M. 2011. The evolving use of advanced high-strength steels for automotive applications. Summer intern report, University of Michigan.
22. Ghali, S. N., H. S. El-Faramawy and M. M. Eissa. 2012. Influence of boron additions on mechanical properties of carbon steel. *Journal of Minerals and Materials Characterization and Engineering* 11: 995–999.
23. Douthit, T. J. and C. J. Van Tyne. 2005. The effect of nitrogen on the cold forging properties of 1020 steel. *Journal of Materials Processing Technology* 160: 335–347.
24. Brauer, H. E. 1920. The properties and heat treatment of high speed steel. Thesis, BSc in Chemical Engineering, College of Liberal Arts and Science, University of Illinois.
25. Reichert, J. M. 2016. Structure and properties of complex transformation products in Nb/Mo-microalloyed steels, PhD dissertation, University of British Columbia, Vancouver, BC.
26. Sureshkumar, P. R., D. R. Pawar and V. Krishnamoorthy. 2011. How to make N2 listen to you in steel making! *International Journal of Scientific & Engineering Research* 2: 1–5.
27. Bhadeshia, H. K. D. H. 2008. Properties of fine-grained steels generated by displacive transformation. *Materials Science and Engineering A* 481–482: 36–39.
28. Llewellyn, D. T. and R. C. Hudd. 2000. *Steels: Metallurgy and Applications*. Butterworth-Heinemann, Oxford.
29. Bhadeshia, H. K. D. H. and S. R. Honeycombe. 2006. *Steels – Microstructure and Properties*. Elsevier, Burlington, MA.
30. Shatynski, S. R. 1979. The thermochemistry of transition metal carbides. *Oxidation of Metals* 13: 105–118.
31. Cahn, R. W. 2007. *Thermo-Mechanical Processing of Metallic Materials*. Elsevier, Oxford.
32. Yamini, S. A. 2008. Effect of titanium additions to low carbon, low manganese steels on sulphide precipitation. PhD dissertation, University of Wollongong, Australia.

33. Vagi, J. J., R. M. Evans and D. C. Martin. 1968. *Welding of Precipitation Hardening Stainless Steels*. Manual for Manufacturing Engineering Laboratory, Battelle Memorial Institute, Columbus, OH.
34. Baker, T. N. 2015. Role of zirconium in microalloyed steels: A review. *Materials Science and Technology* 31: 265–294.

Index